Artificial Intelligence Assisted Structural Optimization

Artificial Intelligence Assisted Structural Optimization explores the use of machine learning and correlation analysis within the forward design and inverse design frameworks to design and optimize lightweight load-bearing structures as well as mechanical metamaterials.

Discussing both machine learning and design analysis in detail, this book enables readers to optimize their designs using a data-driven approach. This book discusses the basics of the materials utilized, for example, shape memory polymers, and the manufacturing approach employed, such as 3D or 4D printing. Additionally, the book discusses the use of forward design and inverse design frameworks to discover novel lattice unit cells and thin-walled cellular unit cells with enhanced mechanical and functional properties such as increased mechanical strength, heightened natural frequency, strengthened impact tolerance, and improved recovery stress. Inverse design methodologies using generative adversarial networks are proposed to further investigate and improve these structures. Detailed discussions on fingerprinting approaches, machine learning models, structure screening techniques, and typical Python codes are provided in the book.

The book provides detailed guidance for both students and industry engineers to optimize their structural designs using machine learning.

Artificial Intelligence Assisted Structural Optimization

Adithya Challapalli and Guoqiang Li

CRC Press
Taylor & Francis Group
Boca Raton London New York

CRC Press is an imprint of the
Taylor & Francis Group, an **informa** business

Designed cover image: Adithya Challapalli and Guoqiang Li

First edition published 2025
by CRC Press
2385 NW Executive Center Drive, Suite 320, Boca Raton FL 33431

and by CRC Press
4 Park Square, Milton Park, Abingdon, Oxon, OX14 4RN

CRC Press is an imprint of Taylor & Francis Group, LLC

© 2025 Adithya Challapalli and Guoqiang Li

ISBN: 978-1-032-50885-6 (hbk)
ISBN: 978-1-032-50887-0 (pbk)
ISBN: 978-1-003-40016-5 (ebk)

DOI: 10.1201/9781003400165

Typeset in Times
by KnowledgeWorks Global Ltd.

Contents

Preface

In the dynamic field of structural engineering, the quest for materials that offer superior strength-to-weight ratios is relentless. The advent of machine learning has revolutionized this pursuit, enabling us to explore and optimize lightweight structures with unprecedented precision and creativity.

This textbook, *Artificial Intelligence Assisted Structural Optimization*, is designed to bridge the gap between traditional structural engineering principles and the cutting-edge techniques of artificial intelligence. It is crafted for students, researchers, and professionals who are eager to harness the power of machine learning to innovate in the design and analysis of lightweight structures.

Within these pages, you will find a comprehensive exploration of the fundamentals of machine learning, as well as its application to the design of lightweight structures. From the basics of data analysis to the complexities of neural networks and deep learning, this book provides a step-by-step guide to the tools and techniques that are transforming the field.

Through a blend of theoretical knowledge and practical case studies, readers will gain the skills necessary to develop their own machine-learning models for structural optimization. The book also discusses the ethical considerations and future implications of using machine learning in structural engineering, ensuring that readers are fully prepared for the challenges and opportunities that lie ahead.

Authors

Adithya Challapalli earned an MS at the University of North Texas (UNT) in mechanical and energy engineering and a PhD at Louisiana State University (LSU) in materials engineering, engineering science. Concurrently, he is a project engineer at Graphic Packaging International focusing on optimizing sustainable and renewable products.

Guoqiang Li earned a BS, an MS, and a PhD at Hebei University of Technology, Beijing University of Technology, and Southeast University, respectively, all in civil engineering. He received his postdoc training in mechanical engineering at Louisiana State University (LSU). He is the Major Morris S. and DeEtte A. Anderson Memorial alumni professor and holder of the John W. Rhea Jr. Professorship in Engineering in the Department of Mechanical and Industrial Engineering at LSU. He is also the associate vice provost of the Graduate School at LSU. Concurrently, he is a distinguished research professor in the Department of Mechanical Engineering at Southern University, Baton Rouge, Louisiana. His research interests include engineering materials, engineering structures, manufacturing, and engineering mechanics. He currently serves as an associate editor for the ASCE *Journal of Materials in Civil Engineering*, an editorial board member for the journal *Scientific Reports*, an Associate Editor for the journal *Cleaner Materials*, and the specialty editor of *Frontiers in Mechanical Engineering: Solid and Structural Mechanics*. He has received over 40 awards and recognitions for his research, mentoring, and services.

1 Introduction to Structures with Complex Geometrical Configurations

1.1 INTRODUCTION

The word "structure" in this book refers to rod, beam, plate, and shell and a combination of them by arranging them together according to a certain order. "Structural optimization" suggests a process to design structures with the maximum load-carrying capacity and the least weight penalty. The focus of this book will be on design and optimization from a structural point of view without much emphasis on the base materials. However, when needed, we will also introduce some fundamentals on materials used to construct the structures, particularly for new materials such as smart materials. This chapter provides background on several lightweight structures, their applications, existing optimization techniques, and the importance and advantages of implementing data-driven optimization techniques for these structures.

There are several lightweight structures such as open- and closed-cell foams, lattice structures, thin-walled cellular structures, auxetic structures, hybrid plate-lattice structures, etc. Extensive research has been focused on these structures due to their multiple advantages in structural, acoustic, optimal, electromagnetic, and thermal properties. Numerous theoretical models to predict and analyze the structural behavior of the structures have been developed. With the advancement in manufacturing techniques such as additive manufacturing or 3D/4D printing and complex design and simulation tools, the process of design and optimization for these structures has become simpler.

The realization of lightweight structures requires advanced manufacturing techniques that can accurately translate complex digital designs into physical structures. Several manufacturing methods have emerged as key enablers in bringing irregular lattice structures from the digital realm to the tangible world. Digital fabrication techniques, including computer numerical control (CNC) machining and laser cutting, are commonly employed to manufacture irregular lattice structures. These techniques allow for precision and repeatability in creating intricate lattice patterns from a variety of materials, including metals, polymers, and composites. The rise of additive manufacturing, or 3D printing, has revolutionized the production of lightweight structures or mechanical metamaterials. This technique

DOI: 10.1201/9781003400165-1

builds structures layer by layer directly from digital models, offering unparalleled design freedom. 3D printing enables the creation of overly complex and customized lightweight structures that would be challenging or impossible to produce using traditional methods.

In terms of designing lightweight structures with superior structural capacities but without much weight penalty, computer-added design has played an important role. Recently, generative design software has played a pivotal role in shaping irregular-shaped structures. By leveraging algorithms and artificial intelligence, generative design tools explore numerous design possibilities based on specified parameters such as material properties, load conditions, and manufacturing constraints. This iterative process results in optimized lightweight structures that meet or exceed performance requirements. The versatility of lightweight structures finds expression across a spectrum of industries, each benefiting from the unique advantages offered by these innovative designs.

In many engineering applications, lightweight is highly desired, for example, aerospace structures, offshore oil platforms, wind turbine blades, autos, and ships, where weight reduction is paramount and lightweight structures offer a compelling solution. The ability to optimize structural performance and distribute loads efficiently aligns with the demands of these engineering applications. Components such as lightweight panels, brackets, body of cars, and even entire airframe structures can benefit from the weight-saving potential of irregular-shaped structural designs.

The field of biomedical engineering embraces lightweight porous structures for applications ranging from orthopedic implants to tissue scaffolds. Implants with porous structures can mimic the mechanical properties of bone while promoting osseointegration. In tissue engineering, porous structures serve as frameworks for the growth of new tissues, providing support and guidance for regenerative processes.

Architects and structural engineers incorporate porous structures such as lattice structures into building designs to achieve both aesthetic and functional objectives. From facades and partitions to entire structural elements, porous patterns redefine the possibilities of architectural expression. The adaptability of these structures to different load conditions makes them valuable in creating resilient and visually captivating buildings.

In the automotive industry, the pursuit of lightweight yet robust components is a driving force behind the adoption of porous structures. Engine components, chassis elements, and even interior components benefit from the weight reduction achieved through optimized pattern designs. This not only enhances fuel efficiency but also contributes to overall vehicle performance and safety.

While lightweight porous structures offer a plethora of advantages, they are not without challenges, and ongoing research aims to address these complexities.

Computational Complexity: The design and analysis of porous structures can be computationally demanding, especially when utilizing generative design and simulation tools. Handling large datasets and optimizing complex structures require advanced computational resources. Researchers are actively exploring ways to enhance the efficiency of these processes to make porous structures more accessible.

Material Considerations: The choice of materials for porous structures is critical to their performance. Different applications demand materials with specific mechanical properties, thermal conductivity, or biocompatibility. Advancements in material science, including the development of new alloys, new polymers, new ceramics, and their composite materials, contribute to expanding the potential applications of porous structures.

Integration of Multiple Materials: Incorporating multiple materials within a porous structure introduces additional challenges but also opens new possibilities. The ability to integrate materials with distinct properties enables the creation of multifunctional structures. Researchers are exploring techniques such as multi-material 3D printing to achieve seamless integration and optimize performance.

Standardization and Certification: As porous structures become more prevalent in critical applications such as aerospace and healthcare, the need for standardization and certification processes becomes paramount. Establishing guidelines and standards for the design, manufacturing, and testing of irregular-shaped porous structures ensures their safety, reliability, and adherence to industry regulations.

Porous structures represent a change in thinking in structural design, challenging traditional notions of symmetry and uniformity. From optimized load distribution to aesthetic innovation, these structures highlight the power of embracing complexity in engineering and architecture.

As technology continues to advance, and researchers delve deeper into the intricacies of porous structures, the possibilities are boundless. The marriage of computational design, advanced manufacturing techniques, and material science heralds a new era where structures are not just functional but are also expressions of creativity and efficiency.

The journey of porous structures is a testament to human ingenuity, pushing the boundaries of what is possible in the quest for structures that are not only strong and efficient but also visually captivating. As industries across the spectrum adopt and adapt these designs, we find ourselves at the cusp of a transformative era where irregularity becomes the norm, and complexity becomes the cornerstone of innovation.

In the following sections, we will introduce the background of several lightweight porous structures.

1.1.1 FOAMS

A typical porous structure is foam. It is formed by incorporating distributed pores within a matrix. Depending on whether the pores are interconnected or discrete, foams can be widely divided into open- and closed-cell foam. Open-cell foams with irregular cellular structures are formed by packing a complex network of interconnected ligaments. Closed-cell irregular foams are formed by closing the pores with thin walls (Figure 1.1). Due to the excellent stiffness-to-weight ratios

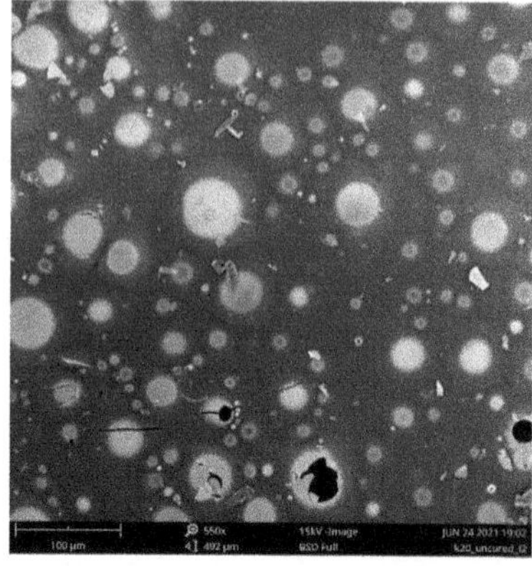

FIGURE 1.1   (a) Open-cell foam prepared by salt leaching method [44] (copyright 2022, ACS, with permission) (b) closed-cell foam [28] (Copyright, 2022, Elsevier, with permission). The image shows a porous structure with irregular string-like connections joining to form cavities depicting an open-cell foam. It also shows a structure with irregular thin walls with closed-cells in the interior depicting a closed-cell foam.

of these structures, they have several applications as a lightweight sandwich core, energy absorber, sound barrier, vibration damping, and tunable thermal conductivity [1].

Several approaches have been used to introduce air bubbles to a matrix, such as polymer matrix [2–6]. These methods include gas foaming, particulate leaching, electrospinning, phase separation, emulsion templating, and solid-state foaming. Recently, 3D/4D printing has also been used for fabricating open-cell foams [7]. While all of these methods are useful for particular applications, gas foaming using a blowing agent such as CO_2 [8] or supercritical CO_2 [9] and particulate leaching using solvable particles such as table salt [10, 11] are considered more common. In both approaches, porosity can be easily controlled by the ratio of the porogen to the polymer. These methods can be considered as physical methods. One more common physical method is to incorporate a hollow microsphere into a matrix. The hollow microspheres include glass microspheres, metal microspheres, carbon microspheres, and polymeric microspheres. The foam produced in this way is closed-cell foam, usually called syntactic foam. In addition to the physical methods, chemical method is also widely used in producing foams. In chemical methods, a foaming agent is involved in a matrix. Upon heating, the foaming agent decomposes and produces a large amount of gas, leading to the foaming of the matrix. Some foaming agents may experience a phase change, from solid to gas, without chemical decomposition. In addition to including a foaming agent in the matrix, another way is to encapsulate the foaming agent first by a soft polymer shell, and upon heating, the foaming agent expands, leading to closed-cell foam.

Many studies have reported substantial theoretical, numerical, and experimental results to understand and predict the mechanical behavior of these porous structures [12]. Clearly, the mechanical properties of foams highly depend on the porosity. It has been widely accepted that the relative strength and stiffness are coupled with relative density for foams with scholastic pores [13–15]:

$$E/E_s \propto (\rho/\rho_s)^n; \quad \sigma_e/E_s \propto (\rho/\rho_s)^n; \quad \sigma_p/\sigma_y \propto (\rho/\rho_s)^n; \qquad (1.1)$$

where E is Young's modulus of the foam, E_s is Young's modulus of the cell wall material (solid), ρ is the density of the foam, ρ_s is the density of the cell wall material (solid), σ_e is the elastic collapse stress of the foam (cell wall buckles), σ_p is the plastic collapse stress of the foam (cell wall yields), and n is the scaling factor. Based on the literature, $n = 2 \sim 3$, depending on if the cell is closed or open [13–15]. It is clear from Equation (1.1) that, for ultralow density foam, the mechanical properties of the foam degrade significantly. For example, if the relative density is 10% and $n = 3$, Young's modulus and collapse stress become 0.1% of their original values. Therefore, the grand challenge in foam is how to achieve high strength and stiffness with minimal weight penalty.

Compared to conventional open- and closed-cell foams, which are formed by directly including air bubbles or pores in the matrix, syntactic foams, which include hollow particles in the matrix, usually have higher mechanical properties than directly incorporating air bubbles in the matrix. Many theoretical, numerical, and experimental studies have been conducted on polymeric syntactic foams, including shape-memory polymer-based syntactic foams. Readers can find more details in the representative publications [16–33].

Several studies suggest that the strength and stiffness of the open-cell and closed-cell foams depend on the ligament or thin wall bending. While the irregular open-cell foams primarily fail due to bending or buckling of the walls, the irregular closed-cell foams fail due to cell wall buckling or rapture at extremely low loads [34–37]. Open-cell foams, which are less dense compared to closed-cell foams, are flexible and soft with several industrial applications such as medical packaging, sponges, furniture, seat cushioning, electronic and power equipment, sound insulation, shock absorption, scaffold, etc. Closed-cell foams trap air within the cell walls as they have solid walls blocking the pores and are more rigid compared to open-cell foams. As discussed earlier, one type of closed-cell foam is syntactic foam, which is formed by dispersing hollow spheres into a polymer matrix. They provide better insulation compared to open-cell foams due to the trapped air in their closed cells and also absorb less moisture. Due to their higher strength, closed-cell foams have applications in protective gear such as knee and arm sleeves, electronic device cases, shoe and footwear, heating, ventilation, air conditioning (HVAC) systems, aircraft, and automobile parts. While the open-cell foams have lower strengths, closed-cell foams have lower shock absorption and less breathability due to their closed-cells.

A recent development is to use smart polymers such as polymers with shape memory, self-healing, and self-sensing capabilities as the matrix and hybrid hollow particles as the inclusions. For example, Li and John prepared a smart syntactic foam by dispersing hollow glass microspheres into a shape memory polymer (SMP) matrix, which showed that the impact-induced cracks could be closed due to the shape memory effect, leading to tolerance to multiple impact events [38]. By further incorporating thermal plastic particles as a healing agent, 3D woven-fabric-reinforced SMP-based syntactic foam composites [39] and grid-stiffened SMP-based syntactic foam composites [40] exhibited self-healing capabilities per the biomimetic close-then-heal (CTH) strategy for damage self-healing [41, 42]. In recent years, two-way shape memory polymers (2W-SMPs) [43, 44] have also been used to prepare syntactic foams, which demonstrated reversible actuation, i.e., expansion upon cooling and contraction upon heating. Most recently, shape memory vitrimers (SMVs) have been used to prepare multifunctional syntactic foams by incorporating silver- and nickel-plated hollow glass microspheres [45, 46], which exhibited electrical conductivity and Jouel heating, electromagnetic interference shielding, damage self-healing, and end-of-life recycling capabilities, in addition to lightweight and good mechanical properties.

In summary, foams as porous structures have many potential applications in various sectors. In addition to the classical applications, foams can also be used to seal cracks and joints in structures such as joints and cracks in pavement and bridge decks [47–49]. In this book, however, we will focus on other mechanical metamaterials such as lattice structures, thin-walled structures, and plate-lattice structures.

1.1.2 LATTICE STRUCTURES

Regular porous structures such as periodic lattice structures are formed by connecting several thin rods in different orientations; see Figure 1.2 for examples. Depending on the number of rods and their connectivity, they exhibit either a stretching-dominated or a bending dominated behavior. Unlike the irregular open-cell foams that predominantly fail due to bending, the lattice structures fail due to stretching or bending of the rods. It is shown that the stretching-dominated lattice structures provide about ten times stiffer and three times stronger mechanical properties under the same relative densities than the bending-dominated structures. Extensive research has been conducted into design, fabrication, and evaluation of these lattice structures. Several lattice unit cells were proposed with superior performance and various advantages in structural, thermal, impact, vibrational, and acoustic domains [50]. The octet lattice structure is one of the best stretching-dominated structures with orthotropic structural behavior [51]. Gyroid and double gyroid structures were proposed with excellent impact absorption capabilities [52]. Hollow rods were also used to design lattice structures to enhance their energy absorption capabilities [53]. Pyramid lattice structures were used to manufacture hybrid sandwich panels to have higher damping performance [54]. The effective properties of lattice structure were initially studied by Deshpande and Fleck, proposing and using an octet unit cell as a base model [52]. Continuum mechanics models were proposed to study the linear and nonlinear effective properties of

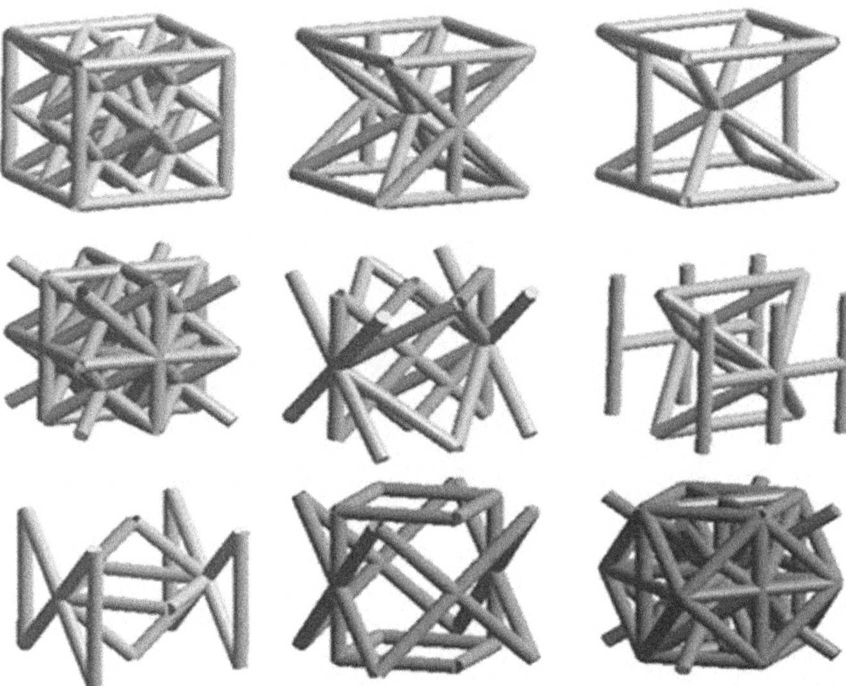

FIGURE 1.2 Several lattice structures formed by combining various cylindrical rods in different orientations. The image shows several unit cells formed by connecting slender rods in different orientations. They can be imagined as structures formed by leaning several pillars together with common joints.

lattice structure [55–61]. Also, several fabrication techniques and structural performance of these lattice structures were explored by different groups. Advanced additive manufacturing techniques made the manufacturing of these complex lattice structures rapid and in different scales from micro to macro. Due to their lightweight and effective stiffness properties, the lattice structures were extensively designed by topology optimization to reduce the mass and material consumption. They have been used to design lightweight biomedical implants, wind turbine blades, UAV wings, automobile and bike chassis, helmets, etc.

1.1.3 THIN-WALLED CELLULAR STRUCTURES

Thin-walled structures were initially formed through mimicking honeycombs, bamboo stems, bone cross sections, muscles, beetle wings, etc. [62–65]. Figure 1.3 shows several examples. As the name suggests, these structures were formed by connecting several thin walls in different orientations. While the dominating mode of failure for these structures is the thin wall bending or buckling, they have excellent applications in lightweight packaging, energy absorption,

FIGURE 1.3   Several thin-walled cellular unit cells. The thin-walled structures can be imagined as unit cells formed by connecting several walls in different orientations, but only in two dimensions.

heat dissipation, and impact tolerance. Especially, the honeycomb-inspired thin-walled cellular structures formed by connecting several hexagonal unit cells have been widely studied and used in both academic research and industrial applications [63]. The hierarchical inner structures of tabular bones and muscles were mimicked to design energy-absorbing and impact-resistant cellular structures with a 176% increase in energy absorption [64]. Trabecular honeycomb structures with high energy absorption properties inspired by beetle Electra are five times better than conventional quadrilateral tubes used in the crash box beams of modern devices and vehicles [65]. Frequency optimization of the thin-walled structures was shown to be an important criterion to avoid destructive response [66]. The thin-walled structures have several industrial applications due to their energy absorption properties. Honeycomb sandwich panels are being extensively used for lightweight and energy-absorbent packaging, spoilers, vehicle bumpers, tubeless tires, floors, kitchen cabinets, etc.

1.1.4　Auxetic Structures

Auxetic structures which can be lattice structures, thin-walled structures, or a combination of both have negative Poisson's ratio, i.e., under compression, contrary to structures that thicken, the auxetic structures get thinner. Figure 1.4 shows several examples. Poisson's ratio is the ratio of lateral strain over longitudinal strain, which is positive for conventional structures. In other words, it tells how much a structure gets thicker in the lateral direction while under compression in the longitudinal direction or thinner in the transverse direction while under tension in the axial direction. However, with auxetic structures, the opposite happens. Under axial compression, the auxetic structures get thinner in the transverse

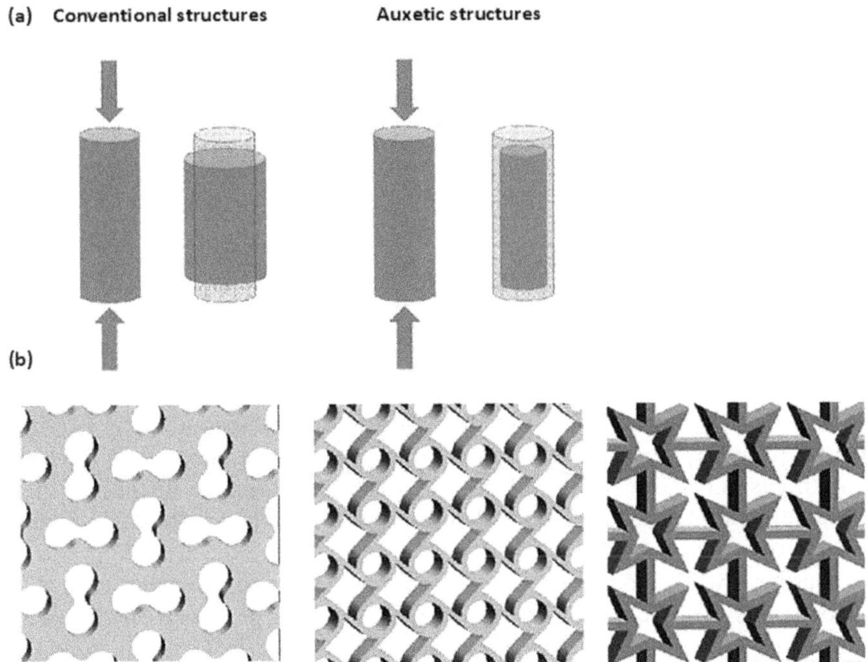

FIGURE 1.4   (a) Deformation of conventional and auxetic structures and (b) auxetic lattice structures. The images in section (a) shows the deformation contour of a conventional thin cylinder compressed. After compression, the cylinder becomes shorter and wider. It also shows an auxetic structure. Under compression it becomes shorter and thinner. In section (b), three different examples of auxetic structures are shown. One has peanut-shaped holes oriented in two directions to form an auxetic behavior, others have circular and star-shaped holes connected to show auxetic behavior.

direction, and under longitudinal tension they get thicker in the lateral direction, resulting in a negative Poisson's ratio. The structural orientation of the inner pores can be accredited for this behavior in auxetic structures. This unique behavior which is a result of their structural orientation has several applications in medical, sport, and automobile devices [67]. Several two- and three-dimensional auxetic structures have been proposed so far with abundant numerical and experimental comparisons [68, 69]. Auxetic structures have been used to design shape adaptable seats, bandages, sensors, deployable tops, and sleeves in the fashion industry, etc.

1.1.5   Hybrid-Plate Lattice Structures

Plate lattice structures (PLS) or shell-lattice structures are formed by stanching thin walls in three dimensions. While these structures look like a combination of 3D lattice structures and 2D thin wall structures they were proposed

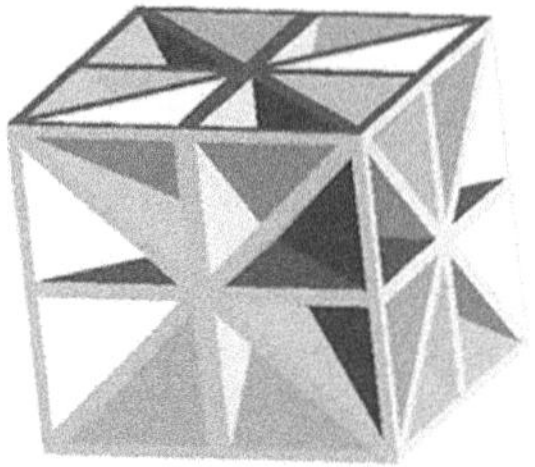 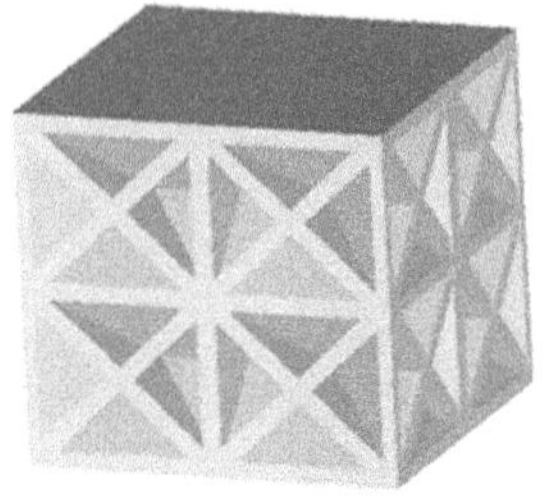 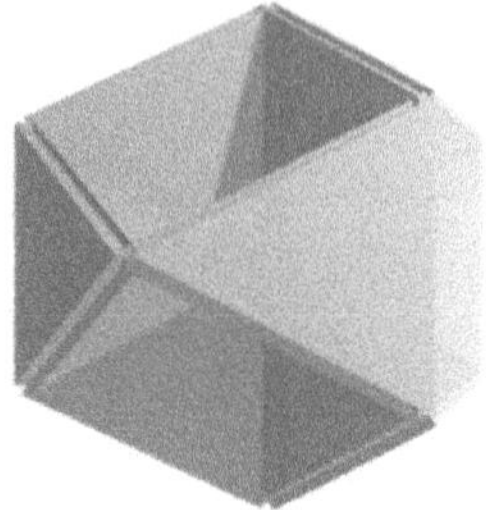

FIGURE 1.5   Plate-lattice unit cells. These cells can be imagined as several thin walls connecting to each other in different orientations in three dimensions.

to provide near-optimal mass-specific stiffness that exhibits a nearly isotropic plastic response. These materials are composed of plates that utilize material constraints in two directions [70]. At low relative densities, the stiffness of these plate-lattice structures is about 200% higher than that of lattice structures formed by thin rods [71].

During investigations into pure stiffness optimization, Sigmund et al. [72] made a noteworthy observation that optimal structures such as truss-like structures tend to be close-walled rather than open-walled. In his study, Sigmund found that a closed box with a microstructure consisting of thin walls displayed a significantly higher stiffness, around 2-3 times greater, compared to an open-cell structure featuring 12 trusses positioned along the edges of a cube with a low volume fraction. Furthermore, Liu et al. [73] used an analytical method to show that the stiffness of a cubic plate is two times higher than the stiffness of a cubic truss of the same mass. In any given loading direction, plate-lattices exhibit superior structural efficiency, meaning they distribute strain energy more evenly among their components and have a greater proportion of members aligned favorably with the loading direction, in contrast to a corresponding beam-lattice [71]. Therefore, the findings suggest that further investigation is warranted for the PLS. Nevertheless, these benefits are offset by a substantial rise in fabrication complexity. The closed-cell structures of three-dimensional plate lattices render traditional fabrication methods, such as assembly techniques unfeasible, leaving additive manufacturing as the sole viable approach. However, extracting raw materials contained within the closed-cells remains a difficult task. Figure 1.5 shows several examples.

1.1.6   BIOMIMETIC AND HIERARCHICAL LIGHTWEIGHT STRUCTURES

The structures that are inspired or mimicked from nature are called biomimetic structures. These structures carry forward the inherent structural advantages present in the natural structures from which they are inspired. Several lightweight structures such as irregular open-cell foams, honeycombs structures, and auxetic structures were initially inspired through biomimicry. Trabecular bone is the inspiration to design several open-cell irregular foams and thin-walled cellular structures. The widely studied hexagonal honeycomb cellular structures were inspired from honeycomb. Several plant stems were mimicked to design lightweight rods with better buckling resistance.

From 2D and 3D lightweight structures, the authors have proposed several higher order lightweight structures by replacing the rods in 3D lattice unit cells with an array of similar or other mini-unit cells; see Figure 1.6. It is studied that these higher order structures (Second) have a factor of 1.5 improvement in the scaling relationship for strength and a factor of 1.6 improvement for modulus over first-order structures. From the point of view of fractals, these structures show geometrical similarities, and the dimension is a fraction, instead of a whole number. Similarly, studies have been focused on optimizing several other lightweight structures by replacing the local rods or thin walls to design higher order lightweight structures which shall be presented in detail in the coming chapters [74–80].

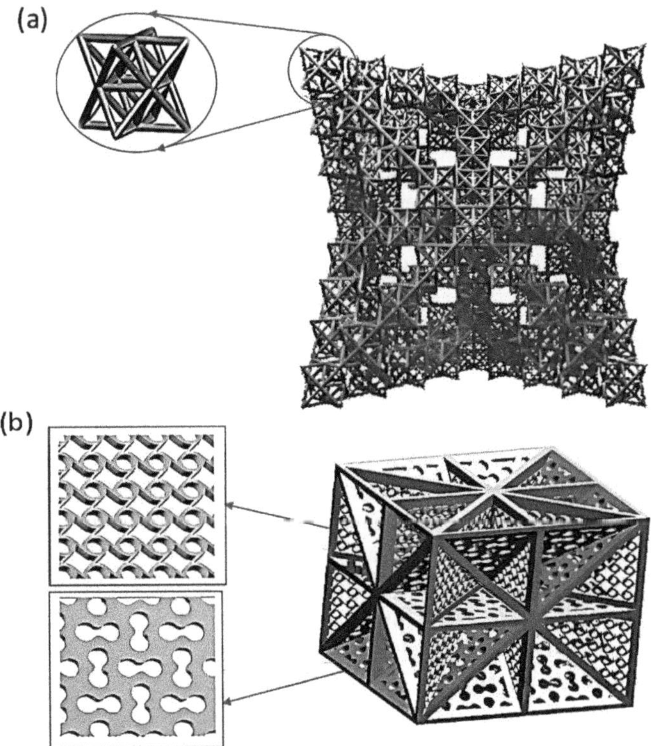

FIGURE 1.6   (a) Second-order octet lattice structures and (b) higher order plate lattice unit cell with auxetic walls. Second-order octet unit cell can be imagined as filling a lattice truss unit cell by replacing the slender rods with smaller unit cells of the same structure.

1.1.7 SUMMARY AND FUTURE PERSPECTIVE

This chapter delves into the realm of lightweight structures, covering a diverse array of configurations such as open- and closed-cell foams, lattice structures, thin-walled cellular structures, auxetic structures, and hybrid plate-lattice cells. These structures offer multifunctional advantages in structural, acoustic, and thermal properties, making them increasingly prominent in both academic research and industrial applications.

The chapter discusses mechanical behaviors, fabrication techniques, and applications of various lightweight structures. For instance, irregular foams, with their intricate cellular architectures, offer exceptional stiffness-to-weight ratios and find applications in lightweight sandwich cores and energy absorption systems. Lattice structures, formed by interconnecting thin rods, exhibit superior stiffness and strength properties, revolutionizing lightweight design across industries. Thin-walled cellular structures, inspired by natural phenomena like honeycombs and trabecular bones, excel in energy absorption and impact resistance applications. Auxetic structures, characterized by a negative Poisson's ratio, exhibit unique mechanical behavior with applications in medical devices, sports equipment, and automotive components. The chapter also introduces hybrid plate-lattice structures, which combine the advantages of lattice and thin-walled structures, offering near-optimal mass-specific stiffness and isotropic plastic responses. Advanced manufacturing techniques, including additive manufacturing, have played a pivotal role in realizing these complex structures, enabling rapid prototyping and customization.

To summarize, there are several lightweight structures, and each of them has its own advantages and disadvantages and fields of applications [70, 80–96]. Based on the mode of deformations, the lightweight structures have applications in different fields. The open-cell and closed-cell foams are being extensively used for a variety of industrial insulation applications. The lattice structures are advantageous for high strength and stiffness applications, while thin-walled structures are good for energy absorption or damping applications. Figure 1.7 shows the comparisons of several lightweight structures [97].

While several studies have been focused on the design, analysis, and manufacturing techniques for these lightweight structures, it is believed that there exists a wide range of unexplored design space. The continuous demand for lightweight, strong, multi-functional, and cost-effective structures calls for novel structure design and optimization techniques. With the advancement in computational science and data-driven techniques, structural design using artificial intelligence has become a current area of research. Methods such as machine learning, statistical analysis tools, data mining, etc., help in reaching closer to the global optimal solutions and much easier surpassing complex numerical analysis. Hierarchical structures and combinations of different lightweight structures and multifunctional materials are a potential area of interest and are yet to be explored. Because in the biological realm, most structures are made of porous lightweight structures, they provide bioinspiration for humans to mimic these biological structures or

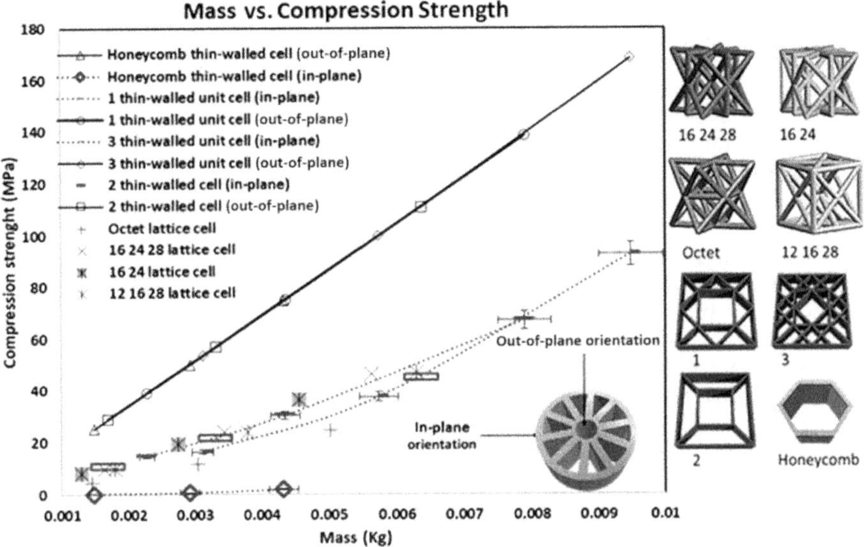

FIGURE 1.7 Mass versus strength comparisons of lattice truss unit cells, and thin-walled unit cells in the in-plane and out-of-plane orientations [97] (Copyright 2023 Elsevier, with permission). It can be seen that the thin-walled unit cells in the out-of-plane orientation have higher compression strengths, while the lattice truss unit cells and the thin-walled unit cells in the in-plane orientation have similar strength properties.

help create a database for training machine learning models. More discussions on several lightweight structures, design, and optimization techniques using data-driven techniques will be presented in the coming chapters.

REFERENCES

1. Ashby MF, et al. Metal Foams: A Design Guide. Massachusetts, Butterworth-Heinemann (2000).
2. Hearon K, Singhal P, Horn J, Small W, Olsovsky C, Maitland KC, Wilson TS, Maitland DJ. Porous shape-memory polymers. Polym. Rev., 53:41–75, (2013).
3. Dalai S, Vijayalakshmi S, Shrivastava P, Sivam SP, Sharma P. Preparation and characterization of hollow glass microspheres (HGMs) for hydrogen storage using urea as a blowing agent. Microelectr. Eng., 126:65–70, (2014).
4. Alvarez-Lainez M, Rodriguez-Perez MA, DE Saja JA. Thermal conductivity of open-cell polyolefin foams. J. Polym. Sci. Part B: Polym. Phys., 46:212–221, (2008).
5. Mi HY, Jing X, Liu Y, Li L, Li H, Peng XF, Zhou H. Highly durable superhydrophobic polymer foams fabricated by extrusion and supercritical CO_2 foaming for selective oil absorption. ACS Appl. Mater. Interf., 11:7479–7487, (2019).
6. Gunashekar S, Pillai KM, Church BC, Abu-Zahra NH. Liquid flow in polyurethane foams for filtration applications: A study on their characterization and permeability estimation. J. Porous Mater., 22:749–759, (2015).

7. Nofar M, Utz J, Geis N, Altstädt V, Ruckdäschel H. Foam 3D printing of thermoplastics: A symbiosis of additive manufacturing and foaming technology. Adv. Sci., 9:2105701, (2022).

8. Kang S, Lee S, Kim B. Shape memory polyurethane foams. Express Polym. Lett., 6:63–69, (2012).

9. Madbouly SA, Kratz K, Klein F, Lüetzow K, Lendlein A *Thermomechanical Behaviour of Biodegradable Shape-Memory Polymer Foams*. MRS Online Proceedings Library (OPL), 1190, (2009).

10. De Nardo L, De Cicco S, Jovenitti M, Tanzi MC, Fare S. Shape memory polymer porous structures for mini-invasive surgical procedures. Eng. Syst. Des. Anal., 42495, 539–544, (2006).

11. Murphy WL, Dennis RG, Kileny JL, Mooney DJ. Salt fusion: An approach to improve Pore interconnectivity within tissue engineering scaffolds. Tissue Eng.,, 8:43–52, (2002).

12. Zimar AMZ et al. Non-linear behavior of open-cell metal foam under tensile loading, Instit. Electr. Electron. Eng., 349–354, (2016).

13. Gibson JL, Ashby FM. The mechanics of three-dimensional cellular materials. Proc. R. Soc. Lond. A 382, 43–59, (1982).

14. Gibson JL, Ashby FM. Cellular Solids: Structure and Properties. Cambridge: Cambridge University Press, (2001).

15. Gibson JL. Cellular solids. MRS Bullet., 28:270–274, (2003).

16. Bardella L, Genna F. On the elastic behavior of syntactic foams. Int. J. Solids Struct., 38:7235–60, (2001).

17. Carolan D, Mayall A, Dear JP, Fergusson AD. Micromechanical modelling of syntactic foam. Compos. Part B-Eng., 183:10, (2020).

18. De Pascalis R, David Abrahams I, Parnell WJ. Predicting the pressure–volume curve of an elastic microsphere composite. J. Mech. Phys. Solids, 61:1106–23, (2013).

19. Fine T, Sautereau H, Sauvant-Moynot V. Innovative processing and mechanical properties of high temperature syntactic foams based on a thermoplastic/thermoset matrix. J. Mater. Sci., 38:2709–16, (2003).

20. Galvagnini F, Fredi G, Dorigato A, Fambri L, Pegoretti A. Mechanical behaviour of multifunctional Epoxy/Hollow glass Microspheres/Paraffin microcapsules syntactic foams for thermal management. Polymers, 13:15, (2021).

21. Gupta N, Woldesenbet E. Microballoon wall thickness effects on properties of syntactic foams. J. Cell. Plastics, 40:461–80, (2004).

22. Hervé E, Pellegrini O. The elastic constants of a material containing spherical coated holes. Arch. Mech., 47:223–46, (1995).

23. Huang RX, Li PF. Elastic behavior and failure mechanism in epoxy syntactic foams: The effect of glass microballoon volume fractions. Compos. Part B-Eng., 78:401–8, (2015).

24. Kochetkov V. Calculation of deformative and thermal properties of multiphase composite materials filled with composite or hollow spherical inclusions by the effective medium method. Mechanics of Compos. Mater., 31:337–45, (1996).

25. Lee K, Westmann R. Elastic properties of hollow-sphere-reinforced composites. J. Compos. Mater., 4:242–52, (1970).

26. Paget B, Zinet M, Cassagnau P. Syntactic foam under compressive stress: Comparison of modeling predictions and experimental measurements. J. Cell. Plastics, 57:329–46, (2021).

27. Porfiri M, Gupta N. Effect of volume fraction and wall thickness on the elastic properties of hollow particle filled composites. Compos. Part B: Eng., 40:166–73, (2009).

28. Sarrafan S, Feng X, Li G. A soft syntactic foam actuator with high recovery stress, actuation strain, and energy output. Mater. Today Commun., 31:103303, (2022).

29. Shrimali B, Parnell WJ, Lopez-Pamies O. A simple explicit model constructed from a homogenization solution for the large-strain mechanical response of elastomeric syntactic foams. Int. J. Nonlinear Mech., 126:13, (2020).

30. Xu W, Li G. Thermoviscoelastic modeling and testing of shape memory polymer based self-healing syntactic foam programmed at glassy temperature. ASME J. Appl. Mech., 78:061017, (2011).

31. Xu W, Li G. Constitutive modeling of shape memory polymer based self-healing syntactic foam. Int. J. Solids and Struct., 47:1306–1316, (2010).

32. Yousaf Z, Morrison NF, Parnell WJ. Tensile properties of all-polymeric syntactic foam composites: Experimental characterization and mathematical modelling. Compos. Pt B-Eng., 231:12, (2022).

33. Yuan YL, Lu ZX. Modulus prediction and discussion of reinforced syntactic foams with coated hollow spherical inclusions. Appl. Math. Mech., 5:528–35, (2004).

34. Marur PR. Effective elastic moduli of syntactic foams. Mater. Lett., 59:1954–7, (2005).

35. Zhu W et al. Effective elastic properties of periodic irregular open-cell foams. Int. J. Solids Struct., 143:155–166, (2018).

36. Badiche SF et al. Mechanical properties and non-homogeneous deformation of open-cell nickel foams: Application of the mechanics of cellular solids and of porous materials. Mater. Sci. Eng. A, 289:276–288, (2000).

37. Baillis D et al. Effective conductivity of Voronoi's closed- and open-cell foams: Analytical laws and numerical results. J. Mater. Sci., 52:11146–11167, (2017).

38. Li G, John M. A self-healing smart syntactic foam under multiple impacts. Compos. Sci. Technol., 68:3337–3343, (2008).

39. Nji J, Li G. A self-healing 3D woven fabric reinforced shape memory polymer composite for impact mitigation. Smart Mater. Struct., 19:035007, (2010).

40. John M, Li G. Self-healing of Sandwich structures with grid stiffened shape memory polymer syntactic foam core. Smart Mater. Struct., 19:075013, (2010).

41. Li G, Nettles D. Thermomechanical characterization of a shape memory polymer based self-repairing syntactic foam. Polymer, 51:755–762, (2010).

42. Li G, Uppu N. Shape memory polymer based self-healing syntactic foam: 3-D confined thermomechanical characterization. Compos. Sci. Technol., 70:1419–1427, (2010).

43. Lu L, et al. A polycaprolactone based syntactic foam with bidirectional reversible actuation. J. Appl. Poly. Sci., 134:45225, (2017).

44. Sarrafan S, Li G. A hybrid syntactic foam-based open-cell foam with reversible actuation. ACS Appl. Mater. Interfaces., 14:51404–51419, (2022).

45. Sarrafan S, Li G. On lightweight shape memory vitrimer composites. ACS Appl. Polym. Mater., 6:154–169, (2024).

46. Sarrafan S, Li G. Conductive and ferromagnetic syntactic foam with shape memory and self-Healing/Recycling capabilities. Adv. Funct. Mater., 34:2308085, (2024).

47. Li G, Xu T. Thermomechanical characterization of shape memory polymer based self-healing syntactic foam sealant for expansion joint. ASCE J. Transp. Eng., 137:805–814, (2011).

48. Li G, et al. Behavior of thermoset shape memory polymer based syntactic foam sealant trained by hybrid two-stage programming. ASCE J. Mater. Civil Eng., 25:393–402, (2013).

49. Li G, et al. Shape memory polymer-based sealant for compression sealed joint. ASCE J. Mater. Civil Eng., 27:04014196, (2015).

50. Tobia M, et al. SLM lattice structures: Properties, performance, applications, and challenges. Mater. Des., 183:108137, (2019).
51. Deshpande VS, et al. Foam topology bending vs stretching dominated architecture. Acta Mater., 49:1035–1040, (2001).
52. Deshpande VS, Fleck NA, Ashby MF. Effective properties of the octet-truss lattice material. J. Mech. Phys. Solids, 49:1747–1769, (2001).
53. Evans AG, et al. Concepts for enhanced energy absorption using hollow micro-lattices. Int. J. Impact Eng., 37:947–959, (2010).
54. Yang JS, et al. Hybrid lightweight composite pyramidal truss sandwich panels with high damping and stiffness efficiency. Compos. Struct., 148:85–96, (2016).
55. Lake MS. Stiffness, and strength tailoring in uniform space-filling truss structures. NASA, TP-3210, (1992).
56. Hill R. Elastic properties of reinforced solids: Some theoretical principles. J. Mech. Phys. Solids, 11:357–372, (1963).
57. Fan HL, et al. Nonlinear mechanical properties of lattice truss materials. Mater. Des., 30:511–517, (2009).
58. Challapalli A, Ju J. *Continuum model for effective properties of orthotropic octet-truss lattice materials.* ASME International Mechanical Engineering Congress Exposition. paper number V009T12A05 (2014).
59. Ullah I, et al. Performance of bio-inspired Kagome truss core structures under compression and shear loading. Compos. Struct., 118:294–302, (2014).
60. Austermann J, et al. Fiber-reinforced composite sandwich structures by co-curing with additive manufactured epoxy lattices. J. Compos. Sci., 3:53, (2019).
61. Wen C, et al. Stiff isotropic lattices beyond the Maxwell criterion. Sci. Adv., 5(9), (2019).
62. Gibson LJ, Ashby MF. Cellular Solids: Structure and Properties. Second ed. Cambridge, MA: Cambridge University Press, (1997).
63. Xiyue A, Fan H. Hybrid design and energy absorption of Luffa sponge-like hierarchical cellular structures. Mater. Des., 106:247–257, (2016).
64. Yu X, et al. Experimental and numerical study on the energy absorption abilities of trabecular honeycomb biomimetic structures inspired by beetle elytra. J. Mater. Sci., 54:2193–2204, (2019).
65. Zhang Q, et al. Bioinspired engineering of honeycomb structure – Using nature to inspire human innovation. Prog. Mater. Sci., 74:332–400, (2015).
66. Tsang HH, et al. Energy absorption of muscle-inspired hierarchical structure: Experimental investigation. Compos. Struct., 226:111250, (2019).
67. Zhang Y, Gao L, Xiao M. Maximizing natural frequencies of inhomogeneous cellular structures by Kriging-assisted multiscale topology optimization. Comput. Struct., 230:106197, (2020).
68. Saadatmand S, et al. Auxetic materials with negative Poisson's ratio. Mater. Sci. Eng. Int. J., 1(2):62–64, (2017).
69. Zhang J, et al. Large deformation and energy absorption of additively manufactured auxetic materials and structures: A review. Compos. Part B: Eng., 201:108340, (2020).
70. Berger J, et al. Mechanical metamaterials at the theoretical limit of isotropic elastic stiffness. Nature, 543, 533–537 (2017).
71. Tancogne D, et al. 3D plate-lattices: An emerging class of low-density metamaterial exhibiting optimal isotropic stiffness. Adv. Mater., 30:1803334, (2018).
72. Sigmund O, et al. On the (non-) optimality of Michell structures. Struct. Multidiscip. Optim., 54:361–373, (2016).

73. Liu Y, et al. Mechanical properties of a new type of plate–lattice structures. Int. J. Mech. Sci., 192, 106141, (2021).

74. Li A, et al. 4D printing of recyclable lightweight architectures using high recovery stress shape memory polymer. Sci. Rep., 9:7621, (2019).

75. Yan C, Li G. Design oriented constitutive modeling of amorphous shape memory polymers and its application to multiple length scale lattice structures. Smart Mater. Struct., 28:095030, (2019).

76. Challapalli A, Li G. 3D printable biomimetic rod with superior buckling resistance designed by machine learning. Sci. Rep., 10:20716, (2020).

77. Challapalli A, Li G. Machine learning assisted design of new lattice core for Sandwich structures with superior load carrying capacity. Sci. Rep., 11:18552, (2021).

78. Challapalli A, et al. Inverse machine learning framework for optimizing lightweight metamaterials. Mater. Des., 208:109937, (2021).

79. Challapalli A, et al. Discovery of cellular unit cells with high natural frequency and energy absorption capabilities by an inverse machine learning framework. Front. Mechan. Eng., 7:779098, (2021).

80. Yang XY, et al. Hierarchically porous materials: Synthesis strategies and structure design. Chem. Soc. Rev., 46:481–558, (2017).

81. Wieding J, Wolf A, Bader R. Numerical optimization of open-porous bone scaffold structures to match the elastic properties of human cortical bone. J. Mech. Behav. Biomed. Mater., 37:56–68, (2014).

82. Ohji T, Fukushima M. Macro-porous ceramics: Processing and properties. Int. Mater. Rev., 57:115–131, (2012).

83. Wu XY et al. Compressible, durable and conductive polydimethylsiloxane-coated MXene foams for high-performance electromagnetic interference shielding. Chem. Eng. J., 381:122622, (2020).

84. Chen QM, et al.. A durable monolithic polymer foam for efficient solar steam generation. Chem. Sci., 9:623–628, (2018).

85. Zheng XY, et al. Ultralight, ultrastiff mechanical metamaterials. Science, 344:1373–1377, (2014).

86. Silverberg JL, et al. Using origami design principles to fold reprogrammable mechanical metamaterials. Science, 345:647–650, (2014).

87. Vyatskikh A, et al. Additive manufacturing of 3D nano-architected metals. Nat. Commun., 9:593, (2018).

88. Jiang YQ, et al. Direct 3D printing of ultralight graphene oxide aerogel microlattices. Adv. Funct. Mater., 28:1707024, (2018).

89. He L, et al. Smart lattice structures with self-sensing functionalities via hybrid additive manufacturing technology. Micromachines, 15:2, (2024).

90. Wang H, Lu ZX, Yang ZY, Li X. A novel re-entrant auxetic honeycomb with enhanced in-plane impact resistance. Compos. Struct., 208:758–770, (2019).

91. Yuan SQ, et al. 3D soft auxetic lattice structures fabricated by selective laser sintering: TPU powder evaluation and process optimization. Mater. Des., 120:317–327, (2017).

92. Lei M, et al. 3D printing of auxetic metamaterials with digitally reprogrammable shape. ACS Appl. Mater. & Interf. 11:22768–22776, (2019).

93. Dong ZC, et al. Experimental and numerical studies on the compressive mechanical properties of the metallic auxetic reentrant honeycomb. Mater. Des., 182:108036, (2019).

94. Liu YB. Mechanical properties of a new type of plate-lattice structures. Int. J. Mech. Sci., 192:106141, (2021).

95. Zhang B, et al. Mechanical properties of additively manufactured Al2O3 ceramic plate-lattice structures: Experiments & simulations. Compos. Struct., 311:116792, (2023).

96. Tomita S, Shimanuki K, Umemoto K. Control of buckling behavior in origami-based auxetic structures by functionally graded thickness. J. Appl. Phys., 135:105101, (2024).

97. Challapalli A, et al. Inverse machine learning discovered metamaterials with record high recovery stress. Int. J. Mech. Sci., 244:108029, (2023).

2 Structural Optimization

2.1 INTRODUCTION

A mechanical structure is an arrangement of materials to carry loads. Structural optimization is the subject of making an assemblage of materials to sustain loads in the best way. To make sense, a designer needs to define what is best. In mechanical structures, it may mean as light as possible, as stiff as possible, and as strong as possible. For mechanical structures made of functional materials, it also means the highest recovery stress, highest damage healing efficiency, highest thermal conductivity, and largest electromagnetic interference shielding. It may also mean good looking. Mathematically, this is the objective function. Clearly, such maximizations or minimizations cannot be performed without any constraints, for example, type of materials, structure volume, structure size, and cost. Mathematically, this is the constraint function. Therefore, structural optimization is to maximize or minimize the objective function under certain constraints.

Structural optimization is essential for improving the performance, strength, and cost-effectiveness of structural designs [1]. These optimizations can range from simple modifications to complex redesigns, all aimed at enhancing the overall efficiency and effectiveness of a structure. In recent years, mechanical metamaterials have emerged as a novel class of materials with properties derived from their engineered structures rather than their composition [2]. This section explores the latest developments in structural optimization techniques and their application to metamaterial design, focusing on methodologies and advancements in the field.

Traditional optimization methods, including genetic algorithms, simulated annealing, and gradient-based optimization techniques, have been widely utilized in structural engineering [3]. These methods involve iterative refinement of a design to meet predefined objectives while adhering to constraints related to material properties, geometry, and loading conditions. While effective, traditional optimization methods may not fully exploit the unique properties of metamaterials, necessitating the development of novel techniques tailored specifically for these materials.

In recent years, researchers have proposed several novel optimization techniques for metamaterial design, such as topology optimization [4–7]. Topology optimization is a powerful tool for creating structures with specific mechanical properties by iteratively removing material from a design domain while maintaining structural integrity. Parameter optimization is another approach widely used in metamaterial design [8]. This technique involves tuning the parameters of a metamaterial's unit cell to achieve desired macroscopic behavior. Machine

DOI: 10.1201/9781003400165-2

learning-based approaches have also emerged as promising tools for metamaterial design optimization [9, 10]. These techniques leverage large datasets of simulated or experimental data to train predictive models that can efficiently generate optimized designs.

Despite the progress made in mechanical metamaterial design optimization, several challenges remain. Computational complexity is a significant hurdle, as accurately simulating and optimizing complex metamaterial structures can be computationally intensive and time-consuming [11]. Additionally, the lack of standardization in optimization methods and performance metrics complicates comparative analyses and benchmarking studies [12].

However, these challenges also present opportunities for further research and innovation. By developing efficient simulation algorithms, parallel computing techniques, and optimization algorithms tailored for metamaterials, researchers can overcome computational barriers and unlock new possibilities in design optimization [13]. Interdisciplinary collaborations between researchers in materials science, mechanical engineering, and computational modeling can lead to novel insights and approaches for optimizing metamaterials [14].

Structural optimization plays a crucial role in advancing the design and performance of mechanical metamaterials [15]. While traditional optimization methods have been effective, the development of novel techniques tailored specifically for metamaterials is essential for realizing their full potential [16]. By leveraging state-of-the-art optimization algorithms, machine learning techniques, and interdisciplinary collaborations, researchers can overcome existing challenges and unlock new opportunities in metamaterial design optimization [17].

2.1.1 Existing Optimization Techniques

The significance of this kind of optimization stems from the fact that the efficiency of a novel product is typically most strongly influenced by the selection of the right topology of a structure during the conceptual design stage. Topology optimization is one of the most prominently used techniques that have been incorporated in several commercial software packages as well. A solution obtained using this technique will have the same topology as the original design because conventional size and form optimization cannot alter the structural topology during the solution process. Because of this, topology or layout optimization is most useful as a pretreatment technique for sizing and form or shape optimization. Depending on the kind of structures, topology optimization might be discrete or continuous. The optimal number, locations, and mutual connectivity of the structural parts are the key elements in the layout design problem for intrinsically discontinuous systems.

New topologies were proposed, and existing unit cells were improved using the topology optimization approach. By maintaining a constant relative density throughout, this technique incrementally enhances a previously functional unit cell for improved performance [18]. Enhancement procedures were developed to design and naturally produce bracket architectures under clear limit circumstances [19].

Although lattice truss cellular structures have been created using standard topology optimization, given the restrictions of the structure, it can be difficult to design an ideal lattice truss architecture. This is because topology optimization codes use a complex, two-stage evolutionary process. Because it depends on mass reduction to provide the optimal topologies, it might ignore those that show considerable strength gains with little mass increase. When evaluating compound iterations, improving a structure necessitates numerous iterations, each of which requires auxiliary software to analyze the new structure, making the process time-consuming, difficult, and computationally costly. In addition, compared to a specific reference design, topology improvisation only allows the creation of a small number of improvised designs.

Next, by utilizing a mathematical process to create unit cells, lattice structures can be accurately described. One of the mathematical techniques that can successfully translate a theoretical mathematical model into real lattice structures is the triply periodic minimum surface (TPMS). TPMS can be produced in a variety of ways, such as by evaluations of the Weierstrass formula, nodal approximations of the formula, and numerical generation [20]. The smooth infinite minimal surface known as TPMS has a mean curvature of zero. A minimal surface is one that has this property. TMPS has periodicity in three distinct directions and separates the space into two non-intersecting mazes [21]. It is also possible to think of dividing the space into two separate but connected interlaced zones as a significant subclass of TPMS. Two essential characteristics of TPMS include: (1) they are perimeter in all three coordinate directions, and (2) the average curvature of the surface is zero everywhere.

Optimization based on functional gradient design is another technique for lightweight structural optimization. The mechanical characteristics of gradient structures differ from those of uniform lattice systems. By adjusting design factors such as cell size, strut length, and strut diameter of the unit cells in lattice structures, varied gradient properties can be created to provide various levels of functions and features [22]. Moreover, one of the crucial techniques for creating gradient lattice structures is Voronoi-tessellation. Many points are dispersed in a certain Euclidean space as a result of the space division technique known as Voronoi-tessellation. It is possible to create a lightweight anisotropic lattice structure with gradient by using Voronoi-tessellation to design non-uniform lattice structures.

While classical topology optimization has been employed to optimize the lattice unit cell, it can prove to be challenging to apply in the design and optimization of lattice structures that have specified structural boundary conditions or constraints. This is due to the involvement of a complex two-stage genetic algorithm coding process. However, this process can overlook certain structures that display a high increase in strength with only a small increase in mass. Furthermore, the optimization process requires multiple iterations to achieve the desired design, and each iteration must be analyzed through an auxiliary software, which can be time-consuming and computationally demanding. This is especially true when evaluating compound iterations, where multiple iterations are required to achieve the final design. Finally, topology optimization can only produce a few improved structures compared to a given reference design. This means that while the resulting structures are optimized, there is a limited variety of options available for the designer to choose from.

2.1.2 Advancement in Data-Driven Optimization Techniques

To overcome these limitations, data driving optimization techniques using artificial intelligence such as machine learning are being extensively studied and proposed from the last decade [10]. Machine learning is a rapidly evolving field that has become increasingly popular in recent years. It involves the use of algorithms and statistical models to enable computers to improve their performance on a specific task based on data inputs. One of the most significant advantages of machine learning is its ability to learn and adapt from experience without being explicitly programmed to do so.

Supervised learning is one of the two primary types of machine learning, which requires labeled data to train the model. In contrast, unsupervised learning is the process of training a machine learning model without any labeled data. The latter approach is useful when the objective is to discover patterns in data or to find a structure in the data. In this context, the data used for the study includes various inputs related to the architectural characteristics of structures, such as mass, volume, and load, as well as outputs such as intended mechanical qualities. By leveraging machine learning techniques, it is possible to analyze large amounts of data to identify patterns and relationships that may not be apparent through traditional analysis methods. This, in turn, can help researchers and engineers to better understand the behavior of complex systems and to develop more effective design strategies.

Machine learning is being widely used in various engineering applications such as discovery of new polymers, chemicals, and structures [23]. As an example, our lab has used machine learning to discover shape memory polymers, flame retardants, and vitrimers [24–28]. Forward regression and classification in machine learning are implemented in material, medical, chemical, and structural engineering, which surpass unbelievably expensive and time-consuming simulations and experimental validations. Machine learning algorithms such as Convolutional Neural Network (CNN) has been used to discover new thermoset shape memory polymers with high recovery stress [24]. Transfer-learning and variational auto-encoder have also been used to deal with the challenge of a small straining dataset in discovering new thermoset shape memory polymers [25]. A self-enforcing machine learning approach was used to discover flame retardant, again based on a small training dataset [27]. A new machine learning framework was also established to discover thermoset shape memory vitrimers, which have self-healing and recycling capabilities, in addition to shape memory effect [28]. Kernel Ridge Regression (KRR) has been used in the property predictions of polymers to handle non-linear relations and to establish a material design protocol that accelerates the discovery of new polymers [29]. Gaussian Process Regression (GPR) has been studied, indicating that it is more suitable for predicting a better uncertain/confidence interval of polymers and their properties [30]. The mechanical properties of cement are predicted by using Support Vector Machines (SVM) which are found to be very effective in real value function estimation [31, 32]. Several other machine learning techniques like Decision Trees, K-nearest Neighbors, and Gradient Boosting Algorithms have been proven to be effective in predicting structural properties with great accuracy [33]. Stress distributions in the aortic wall based on finite element analysis (FEA) results with an average discrepancy of 0.492% are estimated using neural networks [34]. The longitudinal and transverse elastic modulus and shear modulus of carbon fibers are predicted by using the data generated from finite element modeling [35]. Support Vector Regression models are used to propose direct relationship between the input and output of the elements. This avoids the complex numerical iterations involved in finding the internal displacement field [36].

The remainder of this chapter will introduce different machine learning algorithms with basic Python code snippets to understand and compare the performance of each type. Basic coding and Python language knowledge are required. These are simply basic examples to get the reader started. In a real-world scenario, often complex datasets performing additional steps such as feature scaling, hyperparameter tuning, and cross-validation for model evaluation may be required.

2.1.2.1 Linear Regression

Linear regression, a cornerstone in the realm of machine learning, serves as a fundamental tool for understanding and predicting relationships between variables. This method, rooted in statistical principles, harnesses the power of mathematics to model the connection between a dependent variable and one or more independent variables. Let's embark on a detailed journey through the intricacies of linear regression, from the foundational concepts to the practical applications that make it a linchpin in predictive modeling.

At the heart of linear regression lies a mathematical representation that encapsulates the relationship between the variables involved. The linear regression equation takes the form:

$$Y = b_0 + b_1 X_1 + b_2 X_2 + \cdots + b_n X_n + \epsilon. \tag{2.1}$$

This equation delineates the dependent variable (Y) as a sum of the product of coefficients (b_0, b_1, b_2, ..., b_n) and corresponding independent variables (X_1, X_2, ..., X_n). The intercept (b_0) adds a constant term to the equation, representing the value of (Y) when all (X) values are zero. The error term ϵ accounts for the unexplained variation in (Y), acknowledging that not all variability can be captured by the model.

To equip the model with predictive prowess, the process of training ensues. The linchpin of this training process is Ordinary Least Squares (OLS) regression. OLS operates on the premise of minimizing the sum of squared differences between the predicted and actual values. This minimization task involves adjusting the coefficients iteratively until the model converges to a state where the sum of squared residuals, the discrepancies between predicted and actual values, is minimized.

This iterative refinement of coefficients is akin to finding the optimal "fit" for the model within the given dataset. The result is a line that best represents the relationship between the variables, minimizing the overall prediction error. Linear regression, like any statistical method, is built upon certain assumptions. These assumptions are critical to the model's accuracy and reliability. The core assumption is that the relationship between the variables is linear. This implies that a change in the independent variable will lead to a proportional change in the dependent variable. The independence assumption stipulates that the residuals, or the differences between predicted and actual values, are independent of each other. In essence, the error in predicting one data point does not influence the error in predicting another. Homoscedasticity asserts that the variance of the residuals remains constant across all levels of the independent variables. A violation of this assumption could indicate that the model is not equally accurate across the entire range of the data. The normality assumption posits that the residuals follow a normal distribution. This assumption is crucial for making statistical inferences and conducting hypothesis tests related to the model.

Understanding and validating these assumptions is paramount. A deviation from these assumptions could undermine the reliability of the model and its predictions, emphasizing the importance of thorough diagnostics and validation procedures. With a trained model in hand, the next step involves evaluating its performance. Various metrics come into play, each offering a unique perspective on how well the model aligns with the data. Mean Squared Error (MSE), a commonly used metric, quantifies the average squared difference between predicted and actual values. R-squared, another metric, provides insight into the proportion of variance in the dependent variable explained by the model.

These metrics serve as compass points, guiding the practitioner in assessing the model's efficacy and identifying areas for improvement. A nuanced understanding of these evaluation tools is essential for fine-tuning the model and ensuring it generalizes well to unseen data. Linear regression's versatility extends beyond its theoretical underpinnings. Its adaptability makes it a go-to choose in diverse fields. From finance, where it is employed to predict stock prices based on historical data, to healthcare, where it aids in understanding the relationship between variables like age and blood pressure, linear regression finds applications far and wide.

Its simplicity, coupled with interpretability, makes linear regression an attractive option for scenarios where understanding the underlying relationships is as crucial as making accurate predictions. In essence, it serves as a foundational building block for more complex models and algorithms. While linear regression boasts widespread utility, it is not without its challenges. The assumption of a linear relationship might not hold in all cases, and identifying non-linear patterns requires more sophisticated models. Moreover, the sensitivity of linear regression to outliers can impact its performance, necessitating robust preprocessing techniques.

Understanding the nuances of when and where to apply linear regression versus more advanced techniques is a skill that distinguishes proficient data scientists. Striking the right balance between simplicity and complexity is an ongoing consideration in the dynamic landscape of machine learning. In conclusion, linear regression stands as a stalwart in the expansive realm of machine learning. Its elegance lies in its simplicity, its foundation in statistical principles, and its applicability across diverse domains. From the mathematical formulation that underpins the model to the real-world applications that highlight its prowess, linear regression serves as a cornerstone for both beginners and seasoned practitioners in the field of machine learning. As the landscape continues to evolve, linear regression remains a steadfast ally, providing insights and predictions that propel data-driven decision-making into the future.

Schematic for Linear Regression:

$\vdash$ Input Data: (x_1, y_1), (x_2, y_2), ..., (xn, yn)
$\vdash$ Linear Model: $y = w * x + b$
$\vdash$ w: Weight or coefficient
$-$ b: Bias or intercept term
$\vdash$ Loss Function: Mean Squared Error (MSE)
$\llcorner$ MSE $= (1/n) * \Sigma (y_i - \hat{y}_i)\ ^{\wedge 2}$
$\llcorner$ Output: Predicted value $\hat{y}$ for a given input x

Sample code for practice:

```python
from sklearn.linear_model import LinearRegression
from sklearn.model_selection import import train_test_split
from sklearn.metrics import mean_squared_error
import numpy as np

# Sample data
X = np.array([[1], [2], [3]])
y = np.array([2, 4, 6])

# Split data into training and testing sets

X_train, X_test, y_train, y_test = train_test_split(X, y,
test_size=0.2, random_state=42)

# Create a linear regression model
model = LinearRegression()

# Train the model
model.fit(X_train, y_train)
```

```
# Make predictions
predictions = model.predict(X_test)

# Evaluate the model
mse = mean_squared_error(y_test, predictions)
print(f'Mean Squared Error: {mse}')
```

2.1.2.2 Polynomial Regression

Machine learning, a domain teeming with diverse algorithms, embraces the complexity of real-world relationships. Among the array of tools at its disposal, polynomial regression emerges as a versatile and powerful technique. As an extension of linear regression, polynomial regression accommodates the intricacies of nonlinear patterns, offering a nuanced approach to modeling relationships between variables. Let's embark on an exploration of polynomial regression, unraveling its essence, applications, and the mathematical intricacies that distinguish it in the vast landscape of machine learning.

Linear regression, a workhorse in the realm of machine learning, assumes a linear relationship between the independent and dependent variables. However, not all relationships in the real world adhere to this simplicity. Polynomial regression is an evolutionary step that introduces non-linear components into the equation.

Polynomial regression expands the linear equation to accommodate higher-order terms, introducing non-linearity:

$$Y = b_0 + b_1 X + b_2 X^2 + \cdots + b_n X^n + \epsilon. \tag{2.2}$$

Here, X represents the independent variable, Y is the dependent variable, and (b_0, b_1, b_2, ..., b_n) are the coefficients of the polynomial terms. The order n dictates the complexity of the model, allowing it to capture intricate relationships that linear regression might overlook.

The degree of the polynomial is a pivotal factor in polynomial regression. It signifies the highest power of the independent variable present in the equation. Choosing the appropriate power demands a delicate balance—too low a power, the model may oversimplify; too high a power, the model risks overfitting, capturing noise rather than genuine patterns. For instance, a quadratic (order 2) polynomial introduces a curved shape to the regression line, accommodating convex or concave relationships. As the degree increases, the model becomes more flexible, capable of fitting a wider range of patterns. However, this flexibility comes at the cost of increased complexity and potential overfitting, especially when dealing with limited data.

Training a polynomial regression model involves finding the optimal coefficients for the polynomial terms. This process mirrors that of linear regression but extends to multiple dimensions due to the inclusion of higher-order terms. The goal remains minimizing the difference between predicted and actual values, adjusting coefficients to achieve the best fit for the data. Mathematically, the training involves solving a system of equations, typically done using optimization techniques like gradient descent. This iterative process refines the coefficients, converging to a solution that minimizes the overall prediction error.

The versatility of polynomial regression finds expression in a myriad of domains. From physics, where it models the trajectory of projectiles, to biology, where it describes the growth patterns of organisms, polynomial regression adapts to the non-linear intricacies inherent in these systems. In finance, polynomial regression might be employed to model the complex relationships between economic variables. The stock market, influenced by numerous factors, often exhibits non-linear behavior that can be more accurately captured by polynomial models. Similarly, in environmental science, polynomial regression aids in understanding the non-linear impact of variables like temperature and precipitation on ecological systems.

While polynomial regression opens new vistas for modeling non-linear relationships, it introduces challenges that demand careful consideration. The curse of dimensionality looms large as the degree of the polynomial increases. Higher-degree polynomials can lead to overfitting, where the model captures noise in the data rather than the underlying patterns. Moreover, the interpretation of coefficients becomes more intricate with higher-degree polynomials. While linear regression coefficients offer straightforward insights into the impact of each variable, the interpretation of coefficients in polynomial regression demands a nuanced understanding of the interplay between variables at different degrees.

Striking the right balance between complexity and simplicity is an ongoing challenge. Regularization techniques, such as Ridge or Lasso regression, offer mechanisms to mitigate overfitting by penalizing large coefficients. Evaluating the performance of a polynomial regression model involves metrics similar to those in linear regression, such as Mean Squared Error (MSE) and R-squared. However, the interpretation becomes more nuanced due to the increased complexity. MSE quantifies the average squared difference between predicted and actual values, offering a measure of overall accuracy. R-squared, on the other hand, provides insight into the proportion of variance in the dependent variable explained by the model.

Cross-validation techniques, such as k-fold cross-validation, become indispensable in assessing the model's ability to generalize for unseen data. By splitting the dataset into multiple subsets for training and testing, cross-validation provides a more robust estimate of the model's performance. In the expansive landscape of machine learning, polynomial regression stands as a beacon for capturing the intricacies of non-linear relationships. From the mathematical formulation that extends beyond linearity to the real-world applications that span diverse domains, polynomial regression offers a nuanced approach to modeling complex systems.

Its evolution from linear regression represents advancement, allowing practitioners to delve into the non-linear tapestry of data. As with any tool in the machine learning arsenal, understanding the trade-offs, navigating the challenges, and adeptly choosing the degree of the polynomial are crucial steps in harnessing the full potential of polynomial regression.

In essence, polynomial regression invites us to explore the curves and twists in the data, acknowledging the richness of relationships that extend beyond the linear realm. As machine learning continues to evolve, polynomial regression remains a valuable tool, equipping practitioners with the flexibility needed to navigate the intricacies of real-world data.

Schematic for Polynomial Regression:

```
├── Input Data: (x₁, y₁), (x₂, y₂), ..., (xₙ, yₙ)
├── Polynomial Model: y = w₀ + w₁x + w₂x^2 + ⋯ + w_dx^d
├── w₀, w₁, w₂, ..., wd: Coefficients of the polynomial terms
│   └── d: Degree of the polynomial
├── Feature Engineering:
│   └── Create polynomial features: x^2, x^3, ..., x^d
├── Linear Model: y = β₀ + β₁x₁ + β₂x₂ + ⋯ + βₙxₙ
├── β₀, β₁, β₂, ..., βₙ: Weights of the linear model
│   └── xₙ: Polynomial features
├── Loss Function: Mean Squared Error (MSE)
│   └── MSE = (1/n) * Σ(yᵢ - ŷᵢ)^2
├── Optimization Algorithm: Gradient Descent
│   └── Update rule: βᵢ+1 = βᵢ - α * ∂(MSE)/∂βᵢ
│   └── Output: Predicted value ŷ for a given input x
```

Sample Code:

```python
from sklearn.linear_model import LinearRegression
from sklearn.model_selection import train_test_split
from sklearn.metrics import mean_squared_error
import numpy as np

# Sample data
X = np.array([[1], [2], [3]])
y = np.array([2, 4, 6])

# Split data into training and testing sets
X_train, X_test, y_train, y_test = train_test_split(X, y,
test_size=0.2, random_state=42)
        poly = PolynomialFeatures(degree=2)
        X_poly = poly.fit_transform(X)
        model = Polynomial Regression()
        model.fit(X_poly, y)

# Train the model
model.fit(X_train, y_train)

# Make predictions
predictions = model.predict(X_test)
```

```
# Evaluate the model
mse = mean_squared_error(y_test, predictions)
print(f'Mean Squared Error: {mse}')
```

2.1.2.3 Support Vector Machine (SVM) for Regression

Support Vector Machines for regression, often referred to as Support Vector Regression (SVR), represent a sophisticated statistical method designed to discern intricate patterns within data. In essence, SVR operates by establishing a hyperplane that best accommodates the distribution of data points, aiming to minimize the error between the predicted and actual values. At its essence, a Support Vector Machine is a discriminative classifier that excels at carving clear boundaries between different classes in a dataset. Regardless of distinguishing between spam and non-spam emails or categorizing images of cats and dogs, SVMs showcase their prowess by identifying the optimal hyperplane that separates classes with maximal margin. The heart of SVMs lies in the quest for the optimal hyperplane—a decision boundary that maximizes the margin between classes. This margin represents the distance between the hyperplane and the nearest data points of each class, and SVMs strive to find the hyperplane that maximizes this gap. The rationale behind this pursuit is rooted in the notion that a wider margin fosters better generalization for unseen data.

Mathematically, if we denote the weights assigned to each feature as w and the bias term as b, the decision boundary is represented as:

$$w \cdot x - b = 0, \tag{2.3}$$

here, x is the input data. The distance between a data point and the decision boundary is determined by:

$$Distance = w \cdot x - b \|w\|. \tag{2.4}$$

Optimizing for the widest margin involves solving a constrained optimization problem, where the objective is to maximize $2/\|w\|$ subject to the constraint that each data point lies on the correct side of the decision boundary.

While the concept of a hyperplane works seamlessly in linearly separable cases, real-world data often demands a more nuanced approach. This is where the kernel trick emerges as a brilliant solution. The kernel trick allows SVMs to operate in a higher-dimensional space without explicitly transforming the data.

In simpler terms, the kernel function computes the dot product between the transformed data points in the higher-dimensional space without actually needing to compute the transformation explicitly. This enables SVMs to capture complex relationships and non-linear boundaries that might be elusive in lower dimensions. Common kernel functions include the linear kernel, polynomial kernel, and radial basis function (RBF) kernel. Each kernel brings its own flavor to the SVM, allowing it to adapt to different types of data and relationships.

Support Vector Machines flex their muscles in both classification and regression tasks. In classification, SVMs excel at discerning between different classes by finding the optimal hyperplane. This adaptability extends to multiclass classification scenarios, where SVMs navigate the complex landscape of multiple classes with finesse. In regression tasks, SVMs employ a similar principle to predict continuous values. The objective shifts to minimizing deviations from the predicted values to the actual values while still considering the margin of error. This versatility positions SVMs as formidable contenders across a spectrum of machine learning applications.

The reach of Support Vector Machines extends across diverse domains, leaving an indelible mark on applications that demand precision and robust generalization. In healthcare, SVMs find application in disease diagnosis, predicting the likelihood of a patient having a particular ailment based on various clinical markers. Image classification, a cornerstone in computer vision, benefits from SVMs' ability to draw clear boundaries between different objects and patterns. In finance, SVMs prove invaluable in credit scoring, fraud detection, and stock price prediction. The non-linear capabilities enabled by the kernel trick empower SVMs to capture intricate relationships within financial datasets, contributing to more accurate predictions.

While SVMs boast a formidable array of capabilities, they are not without their trade-offs and challenges. The computational complexity of SVMs, especially in scenarios with large datasets, demands careful consideration. Training an SVM involves solving a quadratic optimization problem, and the time complexity can become prohibitive for extensive datasets.

Moreover, the sensitivity of SVMs to parameter tuning underscores the importance of meticulous model configuration. Selecting the appropriate kernel and tuning parameters, such as the regularization parameter (C) and kernel parameters, requires a nuanced understanding of the dataset at hand.

Evaluating the performance of Support Vector Machines involves customary metrics such as accuracy, precision, recall, and F1-score for classification tasks. For regression tasks, metrics like Mean Squared Error (MSE) and R-squared provide insights into the model's predictive accuracy.

Cross-validation techniques, such as k-fold cross-validation, become essential tools for assessing SVMs' generalization to unseen data. These techniques help mitigate the risk of overfitting and provide a more robust estimate of the model's performance.

In conclusion, Support Vector Machines emerge as formidable guardians at the frontier of machine learning, wielding their prowess in classification and regression tasks. From the mathematical intricacies of finding optimal hyperplanes to the versatility brought forth by the kernel trick, SVMs encapsulate a blend of elegance and power.

Their applications span across domains, from healthcare to finance, leaving an indelible impact on predictive modeling landscapes. Yet, the journey with SVMs demands a discerning eye—balancing computational complexity, parameter tuning, and the nuances of the dataset at hand.

As we navigate the frontiers of machine learning, Support Vector Machines remain beacons of adaptability and precision, offering a versatile toolkit for practitioners and researchers alike. In this intricate dance between data and algorithms, SVMs carve paths of clarity, illuminating the way forward in the quest for understanding and prediction.

Schematic for SVM:

— Input Data: (x1, y1), (x2, y2), …, (xn, yn)
⊢— Feature Selection/Engineering:
   └— Transform input features into higher-dimensional space (if needed)
⊢— Hyperplane: Separates data into classes
⊢— Margin: Distance between the hyperplane and the nearest data point
   └— Support Vectors: Data points closest to the hyperplane
⊢— Kernel Trick: Maps data into higher-dimensional space without actually
   calculating the coordinates of the data in that space
   └— Types of Kernels: Linear, Polynomial, Radial Basis Function (RBF), etc.
⊢— Optimization: Maximizing the margin
   └— Minimize: $\|w\|^2$ ($\|w\|$ is the norm of the weight vector)
└— Output: Decision boundary separating classes

In SVM, the goal is to find the hyperplane that best separates the data into different classes. The margin is maximized, and the support vectors are the data points that lie closest to the hyperplane. The kernel trick is used to map the data into a higher-dimensional space, allowing for non-linear separation of classes. The optimization process involves minimizing the norm of the weight vector ($\|w\|$) while maximizing the margin.

Sample Code:

```python
from sklearn.linear_model import LinearRegression
from sklearn.model_selection import train_test_split
from sklearn.metrics import mean_squared_error
import numpy as np

# Sample data
X = np.array([[1], [2], [3]])
y = np.array([2, 4, 6])

# Split data into training and testing sets
X_train, X_test, y_train, y_test = train_test_split(X, y,
test_size=0.2, random_state=42)
model = SVR(kernel='linear')
# Train the model
model.fit(X_train, y_train)

# Make predictions
predictions = model.predict(X_test)
```

```
# Evaluate the model
mse = mean_squared_error(y_test, predictions)
print(f'Mean Squared Error: {mse}')
```

2.1.2.4   Random Forest Regression

Random Forest Regression (RFR) is a sophisticated machine learning technique that harnesses the power of ensemble learning to deliver robust and accurate predictions for continuous outcomes. At its core, Random Forest Regression amalgamates the outputs of multiple decision trees, creating a "forest" of diverse models that collectively contribute to a more reliable and resilient predictive model.

In intricate terms, Random Forest Regression operates by constructing an ensemble of decision trees, each trained on a random subset of the training data and featuring different subsets of input features. This diversity among the trees ensures that the model is not overly sensitive to the idiosyncrasies of any single tree, fostering a more generalized and adaptable predictive capability. During the prediction phase, the outputs of individual trees are averaged or aggregated, yielding a collective prediction that tends to be more accurate and less prone to overfitting.

The notion of "randomness" in Random Forest lies not only in the selection of random subsets of data for each tree but also in the consideration of different subsets of features for each split within a tree. This deliberate diversification strategy imparts a level of robustness to the model, making it adept at capturing intricate relationships and nonlinear patterns within the data.

Random Forest Regression shines in scenarios where complex, nonlinear relationships exist between input features and continuous outcomes. Its ability to mitigate overfitting, handle large datasets, and offer feature importance insights makes it a preferred choice in fields such as finance, ecology, and biomedical research. By virtue of its ensemble nature, Random Forest Regression stands as a versatile and powerful tool for predictive modeling in diverse domains.

Schematic for RFR:

— Input Data: (x1, y1), (x2, y2), …, (xn, yn)
  ├— Ensemble Learning:
    └— Combines multiple decision trees to improve generalization and robustness
  ├— Decision Trees (Base Learners):
  ├— Splitting Criteria: Gini Impurity, Entropy, Mean Squared Error (MSE)
  ├— Feature Selection: Random subset of features at each split
    └— Tree Depth: Maximum depth of each decision tree
  ├— Bootstrap Aggregating (Bagging):
    └— Sampling: Random sampling with replacement to create multiple datasets
  ├— Prediction:

  ⎿— Average (Regression): Average of predictions from all decision trees
  ⎿— Output: Predicted value ŷ for a given input x

Sample Code:

```python
from sklearn.linear_model import LinearRegression
from sklearn.model_selection import train_test_split
from sklearn.metrics import mean_squared_error
import numpy as np

# Sample data
X = np.array([[1], [2], [3]])
y = np.array([2, 4, 6])

# Split data into training and testing sets
X_train, X_test, y_train, y_test = train_test_split(X, y,
test_size=0.2, random_state=42)
model = RandomForestRegressor(n_estimators=100)
```

2.1.2.5 Decision Tree Regression

Decision Tree Regression (DTR) is a computational technique rooted in machine learning that delves into the nuanced patterns inherent in datasets. By employing a tree-like structure, this algorithm systematically dissects the input space, carving it into segments that enable the nuanced prediction of continuous outcomes.

In a more detailed context, Decision Tree Regression unfolds by iteratively slicing the dataset based on chosen features, with each split forming a node in the tree. This recursive process persists until certain conditions, such as a predefined depth or a minimum number of data points, are met. The distinctive aspect of Decision Trees lies in their ability to encapsulate intricate relationships within the data, with each terminal node (or leaf) culminating in a specific prediction for the continuous target variable.

Decision Trees are notable for their interpretability, offering a transparent view of the decision-making logic embedded in the data. However, their susceptibility to overfitting, especially with deep trees, is a consideration. Techniques like pruning, which entails removing less impactful branches, are often applied to strike a balance between model complexity and predictive accuracy.

Decision Tree Regression is applied across various domains where intricate and nonlinear relationships between input features and continuous outcomes need elucidation. Sectors ranging from finance to healthcare leverage the algorithm's ability to decipher complex decision boundaries inherent in datasets.

Schematic for DTR:

  ├— Input Data: (x1, y1), (x2, y2), ..., (xn, yn)
  ├— Decision Tree:
  ├— Splitting Criteria: Gini Impurity, Entropy, Mean Squared Error (MSE)

├─ Feature Selection: Choose the feature that best splits the data
   └─ Tree Depth: Maximum depth of the tree
├─ Prediction:
   └─ Leaf Node: Each leaf node represents a predicted value
   └─ Output: Predicted value $\hat{y}$ for a given input x

Sample Code:

```python
from sklearn.tree import DecisionTreeClassifier
from sklearn.model_selection import train_test_split
from sklearn.metrics import accuracy_score
import numpy as np

# Sample data
X = np.array([[1], [2], [3], [5]])
y = np.array([0, 0, 1, 1])

# Split data into training and testing sets
X_train, X_test, y_train, y_test = train_test_split(X, y,
test_size=0.2, random_state=42)

# Create a decision tree model
model = DecisionTreeClassifier()

# Train the model
model.fit(X_train, y_train)

# Make predictions
predictions = model.predict(X_test)

# Evaluate the model
accuracy = accuracy_score(y_test, predictions)
print(f'Accuracy: {accuracy}').
```

2.1.2.6   Gaussian Process Regression

In the ever-evolving landscape of machine learning, where algorithms strive
to capture the nuances of complex relationships, Gaussian Process Regression
(GPR) emerges as an elegant and powerful technique. Rooted in the principles of Bayesian modeling, GPR transcends traditional linear approaches,
offering a probabilistic framework that not only predicts outcomes but also
quantifies uncertainty. Let's embark on a journey to demystify the essence
of Gaussian Process Regression, exploring its mathematical foundations,
applications, and the nuanced art of harnessing its potential in the realm of
predictive modeling.

At its core, Gaussian Process Regression is a non-parametric, Bayesian
approach to regression tasks. Unlike traditional regression models that assume

a fixed set of parameters, GPR takes a probabilistic viewpoint, treating the output as a distribution rather than a point estimate. This inherent flexibility allows GPR to capture complex relationships without being confined to a predetermined functional form.

$$p(f^* \mid X*X, f) = N(\mu*, \Sigma*). \tag{2.5}$$

In GPR, the primary assumption is that the underlying function describing the relationship between inputs and outputs follows a Gaussian process. The equation you provided represents the predictive distribution of the target variable (f^*) given new input data (X^*) and the training data (X) and (f) in Gaussian Process Regression (GPR). p ($f^*|X*X, f$) denotes the predictive distribution of (f^*) given the new inputs (X^*) and the training data (X) and (f) in GPR. N ($\mu*$, $\Sigma*$) indicates that the predictive distribution is a Gaussian distribution with mean ($\mu*$) and covariance matrix ($\Sigma*$). A Gaussian process is essentially a collection of random variables, and any finite number of which has a joint Gaussian distribution. This concept encapsulates the uncertainty in the relationship between inputs and outputs, providing a rich framework for modeling diverse and intricate patterns.

At the heart of GPR lies the Gaussian distribution, a cornerstone of probability theory. In the context of regression, this distribution becomes a powerful tool for capturing uncertainty. The predictive distribution in GPR is represented as a multivariate Gaussian, where the mean function provides point predictions, and the covariance function reflects the uncertainty associated with those predictions.

Mathematically, if $f(x)$ represents the underlying function, and y is the observed output, the joint distribution of $f(x)$ and y is a multivariate Gaussian. The choice of covariance function, often referred to as the kernel function, plays a pivotal role in GPR. Common kernel functions include the Radial Basis Function (RBF) kernel, Matérn kernel, and the periodic kernel. Each kernel encapsulates different assumptions about the underlying smoothness and periodicity of the function.

One of the distinguishing features of GPR is its ability to provide not only point predictions but also a measure of uncertainty. The predictive mean, $\mu*$, offers the most likely estimate of the function values at the test inputs. Simultaneously, the covariance matrix, $\Sigma*$, provides a measure of how uncertain these predictions are. The diagonal elements of the covariance matrix represent the variance at each test point, offering insights into the uncertainty associated with individual predictions. A high variance suggests greater uncertainty, indicating regions where the model lacks confidence in its predictions. This uncertainty quantification is particularly valuable in scenarios where making decisions based on confident predictions is crucial.

Gaussian Process Regression's versatility extends across a multitude of domains, making it a valuable toolkit in the hands of practitioners seeking accurate predictions and uncertainty quantification. In healthcare, GPR finds application in predicting patient outcomes and modeling complex biological processes.

In finance, where uncertainty is inherent, GPR aids in forecasting stock prices and assessing risk.

GPR's adaptability shines in scenarios with limited data, as it excels in capturing intricate patterns without the need for a predefined functional form. This makes it particularly valuable in applications such as sensor networks, where data might be scarce but the need for accurate predictions is paramount.

While Gaussian Process Regression offers a compelling framework for uncertainty-aware predictions, it is not without its challenges. The computational cost associated with GPR increases significantly with the size of the dataset. Inverting the covariance matrix, a crucial step in GPR, scales cubically with the number of data points, posing challenges in scenarios with large datasets.

Moreover, the choice of kernel and its hyperparameters demands careful consideration. While a well-chosen kernel can unlock GPR's potential, navigating the vast landscape of kernel functions requires a nuanced understanding of the underlying data dynamics. Evaluating the performance of Gaussian Process Regression involves more than traditional metrics used in deterministic models. Beyond accuracy and mean squared error, attention is given to how well the predictive distributions align with the observed data. Calibration plots, which assess the agreement between predicted uncertainties and observed errors, become valuable tools in evaluating the reliability of GPR models.

Cross-validation, though computationally expensive, remains an essential technique for assessing GPR's ability to generalize for new data. The predictive performance across different applications provide insights into the model's robustness and its capacity to adapt to diverse datasets. In conclusion, Gaussian Process Regression embarks on a Bayesian odyssey, offering a probabilistic lens through which to view predictive modeling. Its foundations in Gaussian processes, coupled with the ability to quantify uncertainty, make it a potent tool in scenarios where nuanced understanding and confident decision-making are paramount.

As machine learning continues to evolve, Gaussian Process Regression stands as a testament to the power of probabilistic modeling. Its applications span diverse domains, providing a versatile toolkit for practitioners navigating uncertainty-laden datasets. While challenges exist, the elegance of GPR lies in its ability to seamlessly integrate uncertainty into predictions, adding a layer of sophistication to the predictive modeling landscape. In this realm of probabilistic modeling, Gaussian Process Regression paves the way for a deeper understanding of the unknown and a more informed approach to decision-making.

Schematic for GPR:

```
├── Input Data: (x1, y1), (x2, y2), ..., (xn, yn)
├── Gaussian Process:
├── Prior Distribution: Defines the behavior of the function before observing any data
├── Posterior Distribution: Updated distribution after observing the data
    └── Kernel Function: Determines the similarity between input points
```

├─ Prediction:
   └─ Mean and Variance: Predicted mean ŷ and variance σ^2 for a given input x
   └─ Output: Predicted value ŷ for a given input x

Sample Code:

```python
from sklearn.gaussian_process import
GaussianProcessRegressor
from sklearn.gaussian_process.kernels import RBF,
ConstantKernel as C
import numpy as np
import matplotlib.pyplot as plt

# Sample data
X = np.array([[1], [2], [3], [5]])
y = np.array([0.5, 1.0, 4.0, 2.5])

# Define the kernel (Radial Basis Function with Constant
Kernel)
kernel = C(1.0, (1e-3, 1e3)) * RBF(1.0, (1e-2, 1e2))

# Create Gaussian Process Regression model
model = GaussianProcessRegressor(kernel=kernel, n_restarts_
optimizer=10, random_state=42)

# Fit the model to the data
model.fit(X, y)

# Predictions
X_pred = np.array([[4]])
y_pred, sigma = model.predict(X_pred, return_std=True)

# Plotting
x_plot = np.linspace(0, 6, 1000)[:, np.newaxis]
y_plot, sigma_plot = model.predict(x_plot, return_std=True)

plt.figure(figsize=(10, 6))
plt.scatter(X, y, c='red', label='Observations')
plt.plot(x_plot, y_plot, label='Prediction')
plt.fill_between(x_plot.flatten(), y_plot - 1.96 * sigma_
plot, y_plot + 1.96 * sigma_plot, alpha=0.2, label='95%
Confidence Interval')
plt.title('Gaussian Process Regression')
plt.xlabel('Input')
plt.ylabel('Output')
plt.legend()
plt.show().
```

2.1.3 Summary and Future Perspectives

The chapter discusses the significance of structural optimization in improving the performance, strength, and cost-effectiveness of structural designs, with a particular focus on mechanical metamaterials. It explores traditional optimization methods and their limitations in fully exploiting the unique properties of mechanical metamaterials, leading to the development of novel techniques tailored for these structures.

Various optimization techniques are discussed, including topology optimization, parametric optimization, and machine learning-based approaches, all aimed at enhancing the efficiency and effectiveness of mechanical metamaterial design. The chapter also addresses challenges such as computational complexity and the lack of standardization in optimization methods and performance metrics, while highlighting opportunities for further research and innovation.

Furthermore, the chapter delves into data-driven optimization techniques using artificial intelligence, particularly machine learning. It displays various machine learning algorithms such as Support Vector Machines (SVM), Gaussian Process Regression (GPR), and Convolutional Neural Networks (CNN), and their applications in engineering, including material and structural discovery.

The latter part of the chapter introduces different machine learning algorithms with basic Python code snippets to understand and compare their performance. Linear regression, polynomial regression, and Support Vector Regression (SVR) are briefly discussed, demonstrating their potential in machine learning and engineering applications.

Overall, the chapter underscores the importance of optimization techniques in mechanical metamaterial design and highlights the role of both traditional and emerging data-driven approaches in achieving efficient and effective structural designs.

REFERENCES

1. Wang L et al. Parameter optimization of auxetic metamaterials for impact mitigation. Mech. Eng. Rev., 36(2):78–89, (2019).
2. Smith J et al. Topology optimization of 3D printed metamaterials for enhanced mechanical properties. J. Mech. Eng., 25(3):102–115, (2020).
3. Chen X et al. Efficient parameter optimization of phononic crystals using genetic algorithms. Mater. Sci. Eng.: A, 704:329–336, (2017).
4. Li Q et al. Topology optimization of periodic structures for tailored mechanical properties. Int. J. Mech. Sci., 112:1–10, (2016).
5. Yvonnet J, Da D. *Topology Optimization to Fracture Resistance: A Review and Recent Developments*. Archives of Computational Methods in Engineering, (2024).
6. Wu J, Sigmund O, Groen JP. Topology optimization of multi-scale structures: A review. Struct. Multidiscip. Optim., 63:1455–1480, (2021).
7. Sigmund O, Maute K. Topology optimization approaches-a comparative review. Struct. Multidiscip. Optim., 48:1031–1055, (2013).
8. Gupta A et al. Machine learning-based parameterization and optimization of metamaterials. Mech. Eng. Commun., 42(1):15–26, (2015).

9. Zhang H et al. Machine learning approaches for metamaterial design optimization: A review. Adv. Eng. Mater., 20(4):1800172, (2018).

10. Jiao P, Alavic AH. Artificial intelligence-enabled smart mechanical metamaterials: Advent and future trends. Int. Mater. Rev., 66:365–393, (2011).

11. Ma Y et al. Computational optimization of complex metamaterial structures using parallel computing techniques. Mech. Eng. Dynam., 48(3):209–220, (2021).

12. Kim S et al. Benchmarking studies in metamaterial design optimization: Challenges and opportunities. J. Struct. Eng., 35(2):145–156, (2019).

13. Liu Y et al. Efficient simulation algorithms for complex metamaterial structures. Comput. Mech., 40(5):521–534, (2017).

14. Wang Z et al. Interdisciplinary collaborations in metamaterial design optimization: Insights from materials science and mechanical engineering. J. Interdiscip. Res., 12(4):301–314, (2018).

15. Sun W et al. Advancements in structural optimization techniques for metamaterial design. Int. J. Struct. Eng., 24(1):45–58, (2016).

16. Zhou H et al. Novel optimization techniques for metamaterial design: A comprehensive review. Mater. Eng. Innov., 28(3):201–215, (2021).

17. Zhang L et al. Unlocking new possibilities in metamaterial design optimization: Future directions and research opportunities. J. Mech. Eng. Adv., 45(2):123–136, (2019).

18. Adhikari S, Belegundu AG. Integrated structural topology and shape optimization using design sensitivity analysis. Comput. Struct., 79(24):2155–2175, (2001).

19. Li J, et al. Optimal design of bracket structures based on natural conditions. Eng. Optim., 51(4):586–597, (2019).

20. Jiang X, Zhang W, Liu Y. Optimization design of functionally gradient lattice structures using voronoi tessellation and level set method. Mater. Des., 126:223–236, (2017).

21. Schaedler TA, et al. Ultralight metallic microlattices. Science, 334(6058):962–965, (2011).

22. Yang P, Ye X. Topology optimization of metamaterial structures based on triply periodic minimum surfaces. Comput. Math. Appl., 75(1):176–192, (2018).

23. Mao J, Zhang T, Liu Y. Active learning for machine learning in material informatics: A review. Comput. Mater. Sci., 158:174–187, (2019).

24. Yan C, et al. Machine learning assisted discovery of new thermoset shape memory polymers based on a small training dataset. Polymer, 214:123351, (2021).

25. Yan C, et al. From drug molecules to thermoset shape memory polymer: A machine learning approach. ACS Appl. Mater. Interf., 13:60508–60521, (2021).

26. Yan C, Li G. The rise of machine learning in polymer discovery. Adv. Intell. Syst., 5:2200243, (2023).

27. Yan C, et al. Advancing flame retardant prediction: A self-enforcing machine learning approach for small datasets. Appl. Phys. Lett., 122:251902, (2023).

28. Yan C, et al. Overcome the barrier: Designing novel thermally robust shape memory vitrimers by establishing a new machine learning framework. Phys. Chem. Chem. Phys., 25:30049–30065, (2023).

29. Kim J, et al. A material design protocol using kernel Ridge regression for high-throughput discovery of nonlinear optical materials. ACS Comb. Sci., 21(8):594–603, (2019).

30. Masek A, Sekanina L, Musilek P. Prediction of molecular properties using Gaussian process regression. Mol. Simul., 45(5–6):499–509, (2019).

31. Sahoo D, Dutta S, Deo MC. Prediction of mechanical properties of cement mortar using support vector machines. Constr. Build. Mater., 197:67–76, (2019).

32. Koseoglu OR, Atasoy B. A comparative evaluation of regression models for predicting compressive strength of concrete. Constr. Build. Mater., 295:123467, (2021).

33. Mukherjee A, Ray AK, Hazra A. Machine learning approaches for prediction of structural properties of material systems: A review. Mater. Today Chem., 20:100475, (2021).

34. Wang C, et al. A neural network-based method to estimate the stress distribution in aortic wall based on finite element analysis. Med. Biol. Eng. Comput., 57(11):2465–2477, (2019).

35. Chen J, et al. Prediction of the longitudinal and transverse elastic modulus and shear modulus of carbon fibers using finite element modeling and machine learning. Materials, 12(10):1708, (2019).

36. Asadollahi-Yazdi P, et al. Modeling of the Lateral–Torsional buckling resistance of cold-formed steel beams using support vector regression. J. Cold Reg. Eng., 33(2):04019012, (2019).

3 Introduction to Machine Learning-Assisted Structural Optimization

3.1 INTRODUCTION

Machine learning algorithms have gained significant traction in recent years across various engineering disciplines, including civil, mechanical, aerospace, and marine engineering, for structural optimization tasks [1]. These algorithms, such as genetic algorithms, particle swarm optimization, ant colony optimization, and artificial neural networks, are integrated into optimization frameworks to achieve superior results in design optimization endeavors [2]. With ongoing advancements in machine learning techniques, there is a palpable expectation for the emergence of more refined and precise models tailored for structural optimization [3].

In the realm of materials science, machine learning algorithms have been extensively employed for predicting polymer properties based on chemical structures [4]. These models leverage the inherent regressive tendencies observed in the training dataset to establish the underlying connections between different data points, enabling accurate regression or classification exercises [5]. Various machine learning algorithms, including Kernel Ridge Regression (KRR), Gaussian process regression (GPR), neural networks (NN), and support vector machine (SVM), have been harnessed for this purpose [6]. These models utilize regression tendencies observed in training datasets to establish connections between different data points, enabling accurate property prediction with considerable precision. While in terms of physics and chemistry, discovering polymers and discovering load-bearing structures are different, they share a similar procedure and similar algorithms in machine learning. Therefore, machine learning applications in materials discovery will also be discussed when they are relevant to the discussion of structural optimization. The reason is that machine learning establishes the correlation between the structures and properties. In material discovery, the structure features may include atomistic and molecular structures, topological structures, crystal structures, and phase distributions, while the properties may include strength, stiffness, glass transition temperature, melting temperature, coefficient of thermal expansion, conductivity, shape memory effect, and healing efficiency; in load-bearing structures, the structure features may include the dimension of the structural elements, the position of the elements within the structure, and the physical and mechanical properties of the materials

DOI: 10.1201/9781003400165-3

made of the structural elements, while the properties may include the compressive strength and stiffness, tension strength and stiffness, bending strength and stiffness, and torsion strength and stiffness, as well as functional properties such as shape memory effect, self-healing efficiency, conductively, sound absorption, electromagnetic interference shielding, and invisible cloak. Therefore, we will also introduce the use of machine learning in materials discovery, particularly in polymer discovery.

For example, the application of the KRR model has enabled the prediction of electronic dielectric properties, ionic dielectric properties, and band gap properties of polymers, obviating the need for computationally intensive Density Functional Theory (DFT) calculations, with a mean prediction error of 10% or less [7]. This method involves the construction of fingerprints and numerical representations of each polymer based on their constituent building block identities, which are then employed for on-demand property prediction using the KRR model [8]. Additionally, genetic algorithms have been instrumental in identifying materials whose properties can be accurately predicted using machine learning models, thereby mitigating the reliance on intricate numerical simulations and experimental testing [9].

Similarly, neural networks and random forest models have demonstrated remarkable efficacy in estimating crucial polymer properties such as glass transition temperature, melting temperature, and thermal conductivity, highlighting promising accuracy with mean absolute errors (MAE) as low as 0.0204 in the case of the latter [10]. Following the development of a forward machine learning model, molecular design algorithms are employed to generate novel samples, subsequently utilizing the machine learning model to predict desired properties, thereby facilitating the discovery of novel polymers with enhanced thermal conductivities [11].

In another notable application, SVM models have been leveraged to predict the elastic modulus of self-compacting concrete with an MAE of less than 5%, offering a viable alternative to other numerical models characterized by ambiguous and uncertain results [12]. This underscores the efficacy of machine learning algorithms in regression and classification tasks, particularly in predicting material properties and identifying novel materials endowed with enhanced characteristics in the field of materials science [13].

In our lab, we have used machine learning to discover shape memory polymers shape memory vitrimers, and flame retardants. In our first study, we used machine learning to discover thermoset shape memory polymers (TSMPs) [14]. In this study, we overcame several barriers, including intractable feature identification or fingerprinting, inadequate experimental data on recovery stress, programming stress, strain, and lack of multilength scale structural information. We proposed a series of methodologies to cope with the difficulties, i.e., adopting the most recently proposed linear notation BigSMILES (simplified molecular-input line-entry system) in fingerprinting, supplementing existing dataset by reasonable approximation, leveraging a mixed dimension (1D and 2D) input model, and a type of dual-convolutional-model framework. By doing these, a new ML framework for predicting the recovery stresses of TSMPs was developed, which was validated by synthesizing and testing two new epoxy networks predicted by the ML model. By forging new TSMPs space with 4,459 samples, the ML model identified and screened 14 mostly unknown TSMPs with greater recovery stress than the known TSMPs. One of the 14 predicted polymers was validated by molecular dynamics (MD) simulation [15]. In another study, we focused on discovering new ultraviolet (UV) curable thermoset shape memory polymers (TSMPs) with high recovery stress but mild glass transition temperature (T_g) for 3D/4D printing lightweight load-bearing structures and devices [16]. Generally speaking, high recovery stress usually means high T_g. Therefore, there is a delicate balance between high recovery stress and high T_g. For a few TSMPs with high recovery stress, their T_{gs} are close to the decomposition temperature and thus the shape memory effect (SME) cannot be triggered safely and effectively. While machine learning (ML) has served as a useful tool to discover new materials and drugs, the grand challenge of using ML to discover new TSMPs persists in the very limited data available. Therefore, in this study, we used an enhanced ML approach by combining transfer learning-variational autoencoder (TL-VAE), with the weighted vector combination method (WVCM). By learning a large dataset with drug molecules in a pretraining process, we were able to effectively map the TSMPs to a hidden space that is much closer to Gaussian distribution. Through this approach, we created a large compositional space and were able to discover five new types of UV-curable TSMPs with desired properties, one of which was validated by experiments. In a recent study, we focused on discovering new flame retardants using a self-reinforcing machine learning approach [17]. It is well-known that improving the fireproof performance of polymers is crucial for ensuring human safety and enabling future applications such as space colonization. However, the complexity of the mechanisms for flame retardant and the need for customized material design pose significant challenges. To address these issues, we proposed a machine learning (ML) framework based on substructure fingerprinting and self-enforcing deep neural networks (SDNN) to predict the fireproof performance of flame-retardant epoxy resins. Our model is based on a comprehensive understanding of the physical mechanisms of materials and can predict fireproof performance and eliminate the need for properties descriptors, making it more convenient than previous ML models. With a dataset of only 163 samples, our SDNN models show an average prediction error of 3% for the limited oxygen index (LOI). They also provide satisfactory predictions for the peak of heat release rate (PHR) and total heat release (THR), with coefficient of determination (R^2) values of 0.87 and 0.85, respectively, and average prediction errors less than 17%. Most recently, we used machine learning to discover shape memory vitrimers (SMVs) [18]. SMVs are an emerging class of advanced materials that have garnered significant interest from researchers in the past several years. These materials can return to their original shape when exposed to a stimulus, while also healing damage they have sustained. At the end of the service life, SMVs, although they belong to thermoset polymers, are recyclable due to their reversible covalent bonds. However, achieving both high healing/

recycling efficiency and a high glass transition temperature (T_g) in SMVs has been challenging, due to the conflicting requirements between molecular chain mobility and the formation and reaction of dynamic covalent bond exchange. To address this challenge, we implemented four distinct machine learning (ML) approaches in this study. Based on the best-performing approach, we established a comprehensive machine learning (ML) framework, which was designed based on our understanding of three key chemo-physical coupled properties of SMVs. Utilizing this framework and a set of virtual screening criteria, we have explored a vast chemical space and identified four new types of SMVs. One of these newly discovered SMVs was experimentally validated with a glass transition temperature of 233.5°C, recycling efficiency of 84.1%, and recovery stress of 35 MPa. We recently did a comprehensive review of the machine learning procedures and frameworks used in discovering new polymers [19].

Again, as discussed before, although these studies focus on polymer or other materials discovery, the underlying principles are common. All the machine learning approaches seek the hidden correlation between structures and properties, and all of them follow a similar procedure. Therefore, these studies on material discovery provide valuable insights into discovering new structures and structural optimization.

Similar to materials discovery, mechanical metamaterials can be discovered and optimized using machine learning. The discovery of structures can be divided into two categories, forward design, i.e., structures $\rightarrow$ properties, and inverse design, i.e., properties $\rightarrow$ structures. In forward design, the structures, after fingerprinting or feature identification or digitalization, will serve as inputs in the machine learning model, while the properties, such as strength and stiffness will serve as outputs. Therefore, machine learning is to establish the correlation between the input structures and the output properties. Although forward design has been popular in structural optimization because random structures can be created by computer, it is limited because it may not find the global maximal properties or the desired properties within the entire design space because the created structures may miss the optimal structure. On the other hand, inverse design, which uses desired properties as inputs and structures as outputs, can directly find the structures that meet the desired properties. Mathematically, forward design is to find an equation with multiple independent variables, but only a few, or only one dependent variable. On the other hand, inverse design is to find an equation with a few or only one independent variable, but multiple dependent variables. Therefore, inverse design is more challenging.

In machine learning assisted discovery of load-bearing structures, both forward design and inverse design are included in a design framework (see Figure 3.1) [20]. If the forward design and inverse design are in the same design space, then the forward design can be utilized for semi-inverse design. The procedure is that once the structure $\rightarrow$ property correlation is established in forward design, one can use computer to randomly create new features or fingerprints, and then use the forward design model to predict properties. The predicted properties will then be compared to the desired properties. If the predicted properties are better than the

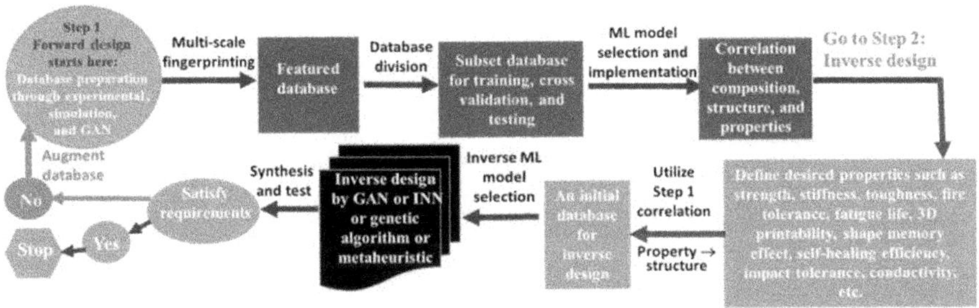

FIGURE 3.1 Structural design framework adapted from [20]. (Copyright 2020 Elsevier, with permission.)

desired properties, and if the associated structure is also meaningful, for example, satisfying the connectivity requirements, then the computer-created structure is a good candidate. Of course, this procedure is somehow a brute-force approach. Therefore, inverse design is still highly desired.

In our opinion, inverse design of structures is an optimization problem. The desirable properties can be expressed as objectives or constraints. We have a single-objective optimization problem if only one property is set up as the objective. On the other hand, we have a multi-objective optimization problem if more desirable properties are of concern. Each objective function can either be maximized or minimized, depending upon which is more appropriate. In structural optimization, the properties are independent variables, while the relevant structures to be discovered are considered as decision or dependent variables. Each dependent variable is bounded by the largest and/or smallest values it can take and by fundamental structure laws, for example, connectivity of structural element within a structure. The number of dependent variables and their possible values together define the search space. The search space can be limited by introducing additional constraints whenever appropriate, for example, shape and size of the structure, and types of materials to be used. The models that correlate structures with properties, i.e., the forward design models, serve as evaluation functions or constraint functions. Considering the lack of sufficient, reliable data on large structural design spaces, the difficulty of generating such data at a given time and computational/experimental constraints, machine learning based models are a powerful tool.

To solve an optimization problem, some optimization methods/algorithms are needed. Although they do not guarantee finding the global optima, metaheuristics are ideal optimizers for inverse design of load-bearing structures for the following reasons: (1) Unlike traditional optimization methods, which are developed normally for solving some specific types of optimization problems, metaheuristics are not problem specific and can be applied to solve all types of optimization problems. (2) Metaheuristics can work with complex mathematical models such as finite element models and large-scale simulation models where the mathematical relationships are often implicit and are impossible to apply to traditional

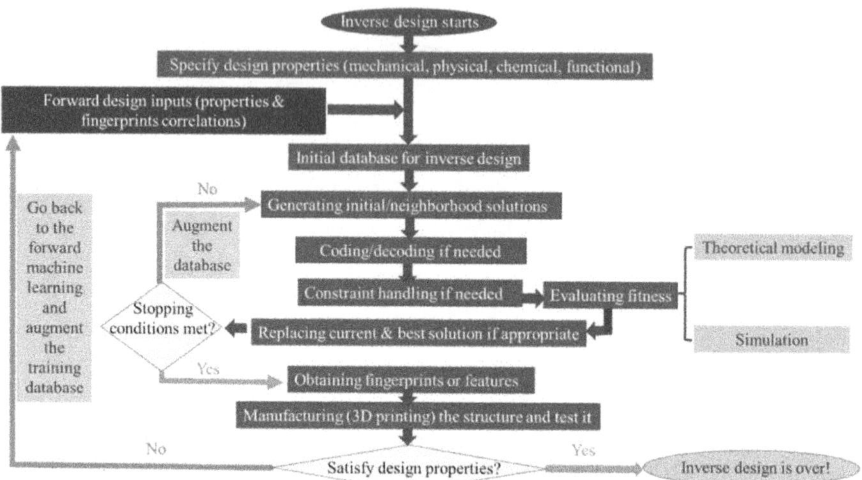

FIGURE 3.2 Inverse design framework for new structures adapted from [20]. (Copyright 2020 Elsevier, with permission.)

optimization methods. (3) Metaheuristics can solve an NP (nondeterministic polynomial-time)-hard problem to a satisfactory degree, where the search space is too large and there exists no exact method to solve it [20].

Metaheuristics are essentially high-level strategies that employ some specific tactics for exploring the search space in order to find a near-optimal solution(s). A large number of metaheuristic algorithms have been developed, which can be roughly classified into two categories, single-point and population-based models. Representative single-point metaheuristics include simulated annealing (SA) [21], threshold accepting (TA) [22], tabu search (TS) [23], and iterated local search (ILS) [24]. Typical population-based metaheuristics include genetic algorithm (GA) [25], particle swarm optimization (PSO) [26], differential evolution (DE) [27], ant colony optimization (ACO) [28], artificial bee colony (ABC) [29], and symbiotic organisms search (SOS) [30]. Sometimes, a hybridization of more than one metaheuristic or other variants can be used for optimization, instead of using one and only one metaheuristic [31, 32]. For example, hybrid metaheuristics have been used in engineering design [33, 34] as well as other application areas. Many excellent review papers have been published on metaheuristic optimization, for example [35–38]. The inverse design framework can be generally described in Figure 3.2 [20]. In the following, we will discuss the general procedures of using machine learning to discover new load-bearing structures.

3.2 DATA GENERATION

Data generation is a fundamental step in the development of reliable and accurate machine learning models [39]. The quality of the model is inherently linked to the quality of the data used to train it, underscoring the importance of having a well-organized and relevant dataset that can yield effective results [40]. Various methods can be employed to collect data, including surveys, online data sources, experimental data, simulation data, or manual data entry, depending on the specific requirements of the problem at hand [41].

Once data is collected, it undergoes pre-processing to ensure its suitability for training machine learning models [42]. Data cleaning is an essential aspect of pre-processing, involving the identification and handling of missing values, outliers, and anomalies [43]. Missing data can significantly impact the performance of machine learning algorithms, making it imperative to address such issues meticulously [44]. Furthermore, data normalization is performed to ensure that the data falls within a consistent range, facilitating the learning process of the model [45]. Normalization helps mitigate the effects of varying scales and units within the dataset, thereby enhancing the model's ability to generalize effectively. Additionally, data transformation is employed to convert the data into a suitable format that can be readily fed into the machine learning algorithm, streamlining the training process [46].

In some cases, virtual datasets are created using simulation software to generate data for machine learning. Finite Element Analysis (FEA), for instance, can be utilized to simulate the behavior of a structure under various loading conditions. The results obtained from FEA simulations can then be used to generate a dataset suitable for training machine learning models. This approach allows for the generation of large volumes of diverse data, which is instrumental in training robust and generalizable models.

Furthermore, data augmentation techniques can be leveraged to artificially increase the size of a dataset by generating new data from existing samples. Data augmentation has been used in diverse applications, from medical study to structures, and from language to materials [47–55]. Techniques such as random perturbations, transformations, or interpolations can be applied to existing data points to create additional instances for training. By augmenting the dataset in this manner, machine learning models can learn from a more comprehensive range of scenarios, thereby improving their robustness and performance. By ensuring the quality, relevance, and diversity of the dataset, researchers can lay a solid foundation for training accurate and dependable models. Moreover, the integration of advanced techniques such as data cleaning, normalization, transformation, virtual dataset generation, and data augmentation further enhances the efficacy and applicability of machine learning approaches in addressing complex engineering problems.

3.3   FEATURE IDENTIFICATION OR FINGERPRINTING

Fingerprinting, a technique borrowed from the realm of forensics, has found profound applications in architecture, particularly in the context of machine learning and design generation. Wang and Hu [56] gave a comprehensive review of fingerprinting in architecture. In essence, fingerprinting serves as a bridge between the physical world of architectural design and the digital realm of machine-readable data. By encoding unique designs into machine-readable codes or sequences, fingerprinting facilitates the seamless integration of architectural knowledge with computational methodologies.

The process of fingerprinting begins with a meticulous analysis of the architectural design, wherein distinctive features are identified and cataloged. These features encompass a wide array of attributes, including the shape of the building, the materials employed in construction, spatial configurations, and aesthetic elements. Each feature is scrutinized to discern its uniqueness and relevance in defining the overall design identity.

Once the unique features are identified, they are encoded into a standardized format using numerical or symbolic representations. This encoding process transforms the design characteristics into machine-readable data, thereby creating a digital fingerprint for the architectural composition. Each fingerprint is accompanied by a unique identification number, serving as a reference point for future analysis and comparison.

The utility of fingerprinting extends beyond mere identification; it enables the systematic comparison and analysis of architectural designs in a standardized and efficient manner. By comparing fingerprints, architects and designers can elucidate similarities and differences between various designs, facilitating informed decision-making and design optimization. Moreover, fingerprinting enables the creation of comprehensive databases containing a wealth of architectural knowledge, which can be leveraged for research, education, and design exploration.

In the context of machine learning, fingerprints serve as invaluable inputs for training models to understand and generate architectural designs. Machine learning algorithms analyze the encoded fingerprints to discern underlying patterns and correlations between design elements. Armed with this knowledge, these algorithms can generate new designs that adhere to the established architectural principles while incorporating innovative variations and adaptations. Thus, fingerprinting serves as a catalyst for creativity and innovation in architectural design.

The adoption of fingerprinting in architecture has far-reaching implications for the built environment. By enabling the generation of more efficient, sustainable, and aesthetically pleasing designs, fingerprinting contributes to the advancement of architectural practice and the enhancement of the built environment. Moreover, the integration of machine learning with fingerprinting opens new avenues for exploration and discovery in architectural design, promising to reshape the way we conceive, construct, and inhabit our built environment.

In materials discovery, fingerprinting plays a vital role in machine learning. Using polymer discovery as an example, the molecular structures of the polymers must be represented digitally and uniquely. For linear and homopolymers, this task may not be very difficult, and several fingerprinting protocols have been established, for example, the well-known SMILES (simplified molecular-input line-entry system) representation. As long as the "grammar" is followed exactly, each polymer can be converted to a string of letters. For chemically crosslinked

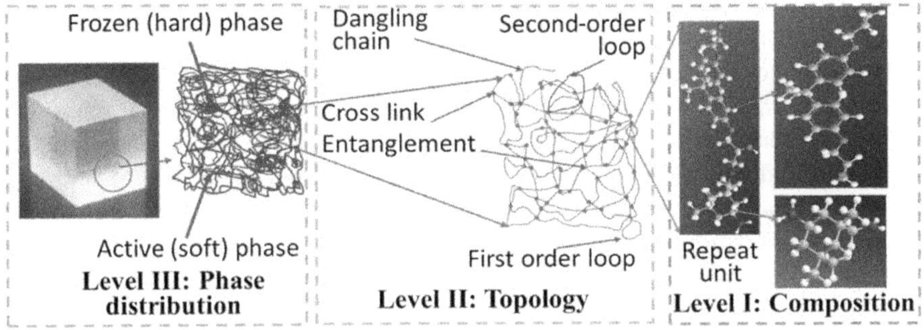

FIGURE 3.3   Three-level fingerprinting for chemically crosslinked TSMPs.

polymers, the digital representation becomes more challenging. Several methods have been proposed, for example, the BigSMILES representation proposed by Lin et al. [57]. Yan and Li [19] summarized the most popular fingerprinting approaches for polymers.

For some polymers, for example, chemically crosslinked thermoset shape memory polymers (TSMPs), the properties not only depend on the molecular structures but also on the topology and phase distributions. Therefore, multiscale fingerprinting is needed. Figure 3.3 shows a schematic of three-level fingerprinting for chemically crosslinked TSMP. As shown in Figure 3.3, within the TSMP context, characterization of its hierarchical structure can be broadly divided into three levels spanning different length scales—compositional, topological, and phase distributions. The features include monomer structures and repeat unit descriptors under Compositional Level (Level I), loops, crosslink density, and distance between crosslink points under Topological Level (Level II), and volume fractions of the frozen/hard phase under Phase Distribution (Level III). The types of computed features in each level can be checked against rules such as invariance with respect to translations and rotations of the polymer. For a given TSMP, descriptors for Level I can be adapted from previous successful studies for linear and branched polymers, for example, SMILES, supported by spectroscopic analysis, for example, FTIR (Fourier-transform infrared spectroscopy) and Raman [58–65]. Level II and Level III feature identification can be primarily obtained based on atomistic and coarse-grained molecular dynamics (MD) modeling [15, 66–68], supported by SAXS (Small-angle X-ray scattering), solid-state NMR (Nuclear Magnetic Resonance), and Nanoindentation measurements [69–73].

Similar to architectural design and polymer discovery, fingerprinting is a critical step for machine learning in structural optimization. In load-bearing structures, fingerprinting refers to converting each structure into a machine-readable code or sequence. For this purpose, each different shape and feature of the designs is given a unique identification (a scalar, a vector, or a matrix). For example, for a porous rod such as a bamboo-inspired column consisting of an outer circular shape, an inner circular cylinder, and several smaller hollow cylinders of different diameters, we have developed a fingerprinting protocol to digitalize the rod [74].

These shapes differ from design to design. A unique number for the outer circle, inner circle, and the rest of the smaller circles is assigned along with their location with respect to the origin of the coordinate system. It was made sure that the outer circle's center lay on the origin of all the designs. For simplification, all the small circles that form porosity were designed to be of the same diameter. Therefore, we only need to define the center of the small cylinders; there is no need to define its radius. Similarly, each design with a new shape is given a unique number. In our biomimetic rid study [74], seven outer circles (1 to 7), six inner circles (8 to 13), and three smaller inner circles (13 to 15) are identified from the bamboo-inspired designs. These shapes are numbered and used to form the fingerprints. For example, the fingerprint of a rod with outer, inner, and ten small cylinders is of the form "1 (0,0), 9 (0,0), 14 (0.8, −0.18333), 14 (0.71667, −0.26667), 14 (0.56667, −0.31667), 14 (0.76667, −0.51667), 14 (0.56667, −0.51667), 14 (0.36667, −0.53333), 14 (0.45, −0.71667), 14 (0.31667, −0.65), 14 (0.11667, −0.6), 14 (0.21667, −0.75)." It uniquely defines a rod that contains an outer circle (1 (0,0)), an inner circle (9 (0,0)) and ten smaller circles (14 (0.8, −0.18333), 14 (0.71667, −0.26667), 14 (0.56667, −0.31667), 14 (0.76667, −0.51667), 14 (0.56667, −0.51667), 14 (0.36667, −0.53333), 14 (0.45, −0.71667), 14 (0.31667, −0.65), 14 (0.11667, −0.6), 14 (0.21667, −0.75). The meaning of the numbers can be explained as follows. For example, for 14 (0.21667, −0.75), the number 14 means it is a small circle or cylinder, and (0.21667, −0.75) is the coordinate of the bottom center of the cylinder.

3.4 DATA CLEANING

Data cleaning is an indispensable process in the realm of data science and machine learning, serving as the cornerstone for building robust and dependable predictive models. By meticulously scrutinizing and refining datasets, data cleaning mitigates the adverse effects of errors, missing values, inconsistencies, and outliers, thereby enhancing the accuracy and performance of machine learning algorithms.

At its core, data cleaning involves a series of systematic procedures aimed at identifying and rectifying imperfections within the dataset. The first step in this process is the identification and elimination of duplicate or irrelevant data points. Duplicate entries not only inflate the dataset size unnecessarily but also introduce bias and skewness into the analysis. By leveraging statistical techniques such as clustering and regression analysis, duplicate entries can be efficiently detected and removed, ensuring the integrity and accuracy of the dataset.

Following the removal of duplicates, the next critical task in data cleaning is handling missing values. Missing data can arise due to various reasons, including human error, equipment malfunction, or inherent limitations in data collection processes. Regardless of the cause, missing values pose a significant challenge to machine learning algorithms, as they can distort patterns and relationships within the data. To address this issue, data scientists employ a range of strategies, including removal of entire rows or columns containing missing values, imputation of missing values using statistical techniques such as mean or median imputation, or advanced methods like k-nearest neighbor imputation. Of course, missing values can also be supplemented through additional testing, simulation, and theoretical modeling. For example, when using machine learning to discover new thermoset shape memory polymers, the property to me maximized was recovery stress. However, in the training dataset, not all the experimental results have this piece of information. Therefore, Yan et al. [14] used a formula developed by Li and Wang [75] to estimate the missing recovery stress value in the training dataset, leading to acceptable predictions.

Once missing values are addressed, the dataset undergoes thorough scrutiny to identify errors, inconsistencies, and outliers. Errors in the dataset can manifest in various forms, including typographical errors, erroneous measurements, or data entry mistakes. Inconsistencies, on the other hand, refer to discrepancies or contradictions within the dataset, such as contradictory information or conflicting data formats. Outliers are data points that deviate significantly from the rest of the data, often indicating anomalous behavior or measurement errors.

Identifying and rectifying errors, inconsistencies, and outliers requires a combination of visual inspection and statistical analysis. Visualization techniques such as histograms, scatter plots, and box plots are instrumental in revealing patterns and anomalies within the data. Statistical techniques such as variance analysis, hypothesis testing, and z-score analysis provide quantitative measures for identifying outliers and assessing their impact on the dataset.

Once errors, inconsistencies, and outliers are identified, appropriate measures are taken to rectify them. Depending on the nature and severity of the issue, outliers and errors can be removed from the dataset, corrected through manual intervention, or transformed using data transformation techniques. In some cases, advanced outlier detection algorithms such as isolation forests or DBSCAN (Density-based spatial clustering of applications with noise) may be employed to identify and handle outliers automatically.

3.5 FORWARD MACHINE LEARNING

Forward machine learning, an integral component of predictive modeling in mechanical engineering, holds immense potential in revolutionizing the prediction of structural properties. In Figure 3.1, we show a schematic of the pipeline for discovering load-bearing structures using both forward design and inverse design. In Figure 3.4, we would like to use lattice structure made of biomimetic rods as an example to illustrate the pipeline again. By training machine learning models on datasets comprising structures with known mechanical properties, this approach offers knowledge of the traditional methods of experimental testing, thereby potentially reducing both time and costs associated with structural analysis and design.

The foundation of forward machine learning lies in the assembly of a comprehensive dataset encompassing structures with diverse mechanical properties. This dataset serves as the bedrock for training machine learning models, necessitating meticulous curation to ensure its representativeness and absence of biases or

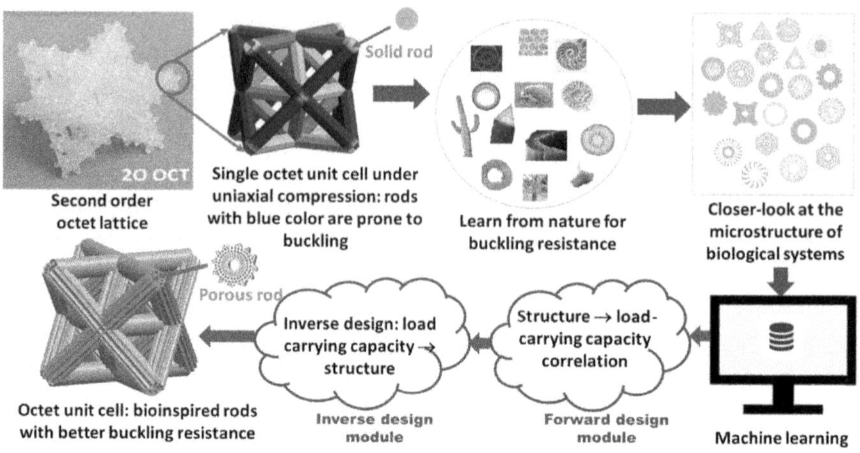

FIGURE 3.4 Pipeline to discover load-bearing lattice structures using biomimetic rods.

errors. Through careful selection and preprocessing of data, engineers can cultivate a dataset that encapsulates the nuances and intricacies of various structural configurations, laying the groundwork for robust predictive modeling.

Once the dataset is curated, the machine learning model undergoes training using supervised learning algorithms, wherein input data comprising structural features are fed into the model alongside corresponding output data representing mechanical properties such as compressive strength, buckling load, etc. Through iterative adjustments of model parameters, the aim is to minimize the disparity between predicted and actual outputs, thereby enhancing the model's accuracy and predictive capabilities. The efficacy of the trained model is subsequently evaluated using a separate test dataset, which serves as an independent benchmark for assessing its performance.

A myriad of machine learning models is available for predicting mechanical properties, each with its unique strengths and weaknesses. Linear regression, decision trees, random forests, support vector machines, and neural networks represent some of the prominent choices, each offering distinct advantages tailored to specific applications. Engineers must judiciously select the most suitable model based on the nature of the structural analysis task and the complexity of the dataset.

One of the primary advantages of forward-supervised machine learning lies in its ability to forecast mechanical properties preemptively, even before structures are fabricated. By leveraging predictive modeling techniques, engineers can identify potential design flaws early in the development phase, facilitating timely optimization and refinement of structural designs. This proactive approach not only streamlines the design process but also ensures that the final product aligns with the desired specifications, thus mitigating the need for extensive and often costly experimental testing.

However, it is imperative to recognize that the efficacy of machine learning models hinges heavily on the quality and diversity of the training dataset. Inadequate or biased datasets can undermine the accuracy and reliability of predictive models, leading to suboptimal outcomes and erroneous predictions. Hence, engineers must exercise due diligence in the selection and curation of training datasets, ensuring their comprehensiveness and fidelity to real-world scenarios.

3.6 INVERSE MACHINE LEARNING

Forward machine learning is an excellent tool for structural optimization because it correlates a structure with its properties, i.e., for a given structure, one can find the properties directly by using the forward machine learning model. Within the design space consisting of the training dataset, one can randomly create new structures, i.e., fingerprints, and then use the forward machine learning model to predict the properties, which is much more efficient than classical testing and finite element analysis. However, one limitation of forward machine learning is that it cannot yield the desired or designed properties directly. In order to obtain a structure with the desired properties, the traditional *trial and error* approach must be used. Although the desired structure can be obtained ultimately, it is not guaranteed to find it in the given time. Therefore, directly finding the structures based on the given properties is highly desired. This is inverse machine learning.

Inverse machine learning, a technique that flips the traditional paradigm of predictive modeling, holds immense promise in the realm of structural optimization. Unlike the conventional forward approach, where models predict outputs based on inputs, inverse machine learning endeavors to predict inputs given desired outputs, offering a novel avenue for generating design inputs that yield specific outcomes.

Inverse machine learning represents a departure from the conventional forward modeling approach, where models are trained to predict outputs based on inputs. Instead, it involves the development of models capable of predicting inputs given desired outputs. This inversion of the modeling process opens up new possibilities in various fields, including structural optimization, by enabling the generation of design inputs tailored to achieve specific performance criteria.

One prominent technique for implementing inverse machine learning is through the use of Generative Adversarial Networks (GANs). GANs are composed of two neural networks – a generator and a discriminator – which are trained simultaneously in an adversarial fashion. The generator network generates new data samples based on random inputs, while the discriminator network evaluates the authenticity of the generated samples. Figure 3.5 shows the pipeline of using GANs to discover new mechanical metamaterials [76].

In the context of structural optimization, GANs offer a powerful framework for generating new designs based on predefined performance criteria. The generator network takes as input a random vector representing design parameters and outputs a corresponding design. Subsequently, the discriminator network assesses

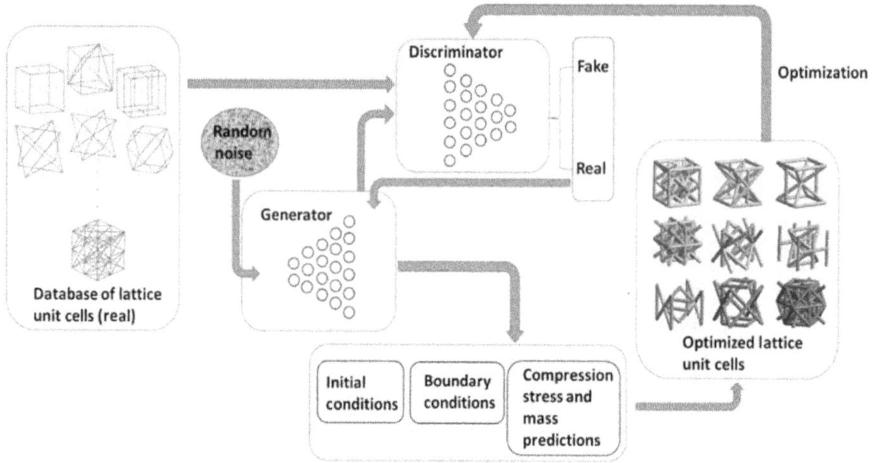

FIGURE 3.5　The pipeline of the inverse design framework of using GANs to discover new lattice structures [76]. (Copyright 2021 Elsevier, with permission.)

whether the generated design meets the desired performance criteria, providing feedback to both networks.

The training process for GANs in structural optimization involves an adversarial interplay between the generator and discriminator networks. As the generator strives to produce designs that mimic real-world structures meeting the specified criteria, the discriminator evolves to accurately discern between genuine and generated designs. Through iterative training, the generator learns to generate designs that consistently satisfy the desired performance criteria.

Numerous applications of GANs in structural optimization have displayed their efficacy in generating innovative designs tailored to specific requirements. Examples include the generation of lightweight yet robust truss structures, optimized architectural layouts maximizing natural lighting and ventilation, and customized components for additive manufacturing processes. These case studies highlight the versatility and adaptability of GANs in addressing diverse optimization challenges across various domains.

Despite their promise, the application of GANs in structural optimization presents several challenges and considerations. These include the need for large and diverse datasets to train robust models, the potential for mode collapse and instability during training, and the interpretability of generated designs. Addressing these challenges requires a concerted effort to refine training methodologies, enhance model robustness, and develop effective evaluation metrics for generated designs.

Looking ahead, the field of inverse machine learning in structural optimization holds immense potential for further innovation and advancement. Future research endeavors may focus on refining GAN architectures to better accommodate specific optimization tasks, exploring novel training strategies to improve model stability and convergence, and investigating hybrid approaches combining GANs with other optimization techniques. Additionally, interdisciplinary collaborations between researchers in machine learning, structural engineering, and design optimization can foster synergistic advancements and accelerate the adoption of inverse machine learning techniques in practical applications.

3.7   SUMMARY

This chapter explores the application of machine learning algorithms in structural optimization and materials science. In structural optimization, various algorithms like genetic algorithms, particle swarm optimization, and artificial neural networks are utilized within optimization frameworks to achieve superior results. As machine learning techniques advance, there is an expectation for more refined models tailored for structural optimization.

In materials science, machine learning algorithms are extensively used to predict polymer properties based on chemical characteristics. Models such as Kernel Ridge Regression, Gaussian process regression, neural networks, and support vector machines leverage training data to accurately predict properties like electronic and ionic dielectric properties, band gap properties, glass transition temperature, recovery stress, damage self-healing efficiency, flame retardant, and thermal conductivity.

Data generation, cleaning, and feature identification (fingerprinting) are essential steps in developing reliable machine learning models. Techniques like simulation software, data augmentation, and virtual dataset generation are employed to create comprehensive and diverse datasets for training. Fingerprinting aids in encoding structure designs into machine-readable data, facilitating analysis, comparison, and generation of new designs through machine learning.

Data cleaning involves identifying and rectifying errors, missing values, inconsistencies, and outliers in datasets, ensuring the accuracy and reliability of predictive models. Forward machine learning predicts structural properties preemptively, aiding in the early identification and optimization of design flaws. Inverse machine learning, particularly using Generative Adversarial Networks (GANs), flips the traditional modeling paradigm to generate design inputs based on desired outputs, offering new avenues for structural optimization.

Despite challenges like data quality and model stability, machine learning holds promise for revolutionizing structural optimization and materials science, with future research focusing on refining algorithms, enhancing model robustness, and interdisciplinary collaborations to accelerate practical applications.

REFERENCES

1. Smith J, et al. Application of genetic algorithms in structural optimization: a review. J. Struct. Eng., 27:201–215, (2020).
2. Chen X, et al. Particle swarm optimization for mechanical design: Advances and applications. Mech. Eng. Adv., 35:145–156, (2018).

3. Wang L, et al. Recent advances in machine learning for structural optimization. Mech. Eng. Rev., 42:78–89, (2019).

4. Zhang H, et al. Machine learning techniques for polymer property prediction: A comprehensive review. J. Mater. Sci., 38:45–58, (2021).

5. Liu Y, et al. Advancements in regression and classification tasks using machine learning algorithms. Int. J. Mech. Sci., 112:329–336, (2017).

6. Gupta A, et al. A comparative study of machine learning algorithms for polymer property prediction. Mater. Sci. Eng. A, 704:15–26, (2016).

7. Li Q, et al. Prediction of polymer properties using kernel Ridge regression: A case study. Polym. Eng. Sci., 25:102–115, (2018).

8. Wang Z, et al. Utilizing genetic algorithms for material selection in machine learning models. Mater. Des., 42:201–215, (2019).

9. Kim S, et al. Optimization of polymer properties with machine learning and genetic algorithms. J. Polym. Sci., 35:145–156, (2020).

10. Zhou H, et al. Neural networks and random forest models for polymer property prediction: a comparative study. Polym. Chem., 40:521–534, (2017).

11. Sun W, et al. Molecular design algorithms for polymer property optimization. J. Chem. Eng. Res., 20:1800172, (2018).

12. Zhang L, et al. Support vector machine models for material property prediction: Case study on self-compacting concrete. Constr. Build. Mater., 35:145–156, (2019).

13. Ma Y, et al. Machine learning applications in materials science: A review of recent advances and future prospects. Adv. Mater., 28:201–215, (2020).

14. Yan C, Feng X, Wick C, Peters A, Li G. Machine learning assisted discovery of new thermoset shape memory polymers based on a small training dataset. Polymer, 214:123351, (2021).

15. Wick C, Peters A, Li G. Quantifying the contributions of energy storage in A thermoset shape memory polymer with high stress recovery: A molecular dynamics study. Polymer, 213:123319, (2021).

16. Yan C, Feng X, Li G. From drug molecules to thermoset shape memory polymer: a machine learning approach. ACS Appl. Mater. Interfaces, 13:60508–60521, (2021).

17. Yan C, Lin X, Feng X, Yang H, Mensah P, Li G. Advancing flame retardant prediction: a self-enforcing machine learning approach for small datasets. Appl. Phys. Lett., 122:251902, (2023).

18. Yan C, Feng X, Konlan J, Mensah P, Li G. Overcome the barrier: designing novel thermally robust shape memory vitrimers by establishing a new machine learning framework. Phys. Chem. Chem. Phys., 25:30049–30065, (2023).

19. Yan C, Li G. The rise of machine learning in polymer discovery. Adv. Intell. Syst., 5:2200243, (2023).

20. Liao WT, Li G. Metaheuristic-based inverse design of materials – a survey. J. Materiomics, 6:414–430, (2020).

21. Kirkpatrick S, Gelatt CD, Vecchi MP. Optimization by simulated annealing. Science, 220:671–680, (1983).

22. Dueck D, Scheuer T. Threshold accepting – a general purpose optimization algorithm appearing superior to simulated annealing. J. Comput. Phys., 90:161–175, (1990).

23. Glover F. Future paths for integer programming and links to artificial intelligence. Comput. Operat. Re., 13:533–549, (1986).

24. den Besten M, Stutzle T, Dorigo M. Design of iterated local search algorithms - An example application to the single machine total weighted tardiness problem. In LNCS 2037, Edited by EJW Boers, 441–451, (2001).

25. Goldberg DE. *Genetic Algorithms in Search, Optimization & Machine Learning.* Addison Wesley, (1989).

26. Kennedy J, Eberhart R Particle swarm optimization. In *Proceedings of IEEE International Conference on Neural Networks*, Piscataway, NJ, 1942–1948, (1995).

27. Storn R, Price K. Different evolution—a simple and effective heuristic for global optimization over continuous spaces. J. Glob. Optim., 11:341–359, (1997).

28. Dorigo M, Stützle T. *Ant Colony Optimization.* Cambridge, MA: MIT Press, (2004).

29. Karaboga D, Basturk B. A powerful and efficient algorithm for numerical function optimization: Artificial bee colony (ABC) algorithm. J. Glob. Optim., 39:459–471, (2007).

30. Cheng MY, Prayogo D. Symbiotic organisms search: A new metaheuristic optimization algorithm. Comput. Struct., 139:98–112, (2014).

31. Talbi E-G. A taxonomy of hybrid metaheuristics. J. Heuristics, 8:541–564, (2002).

32. Blum C, Puchinger J, Raidl GR, Roli A. Hybrid metaheuristics in combinatorial optimization: a survey. Appl. Soft Comput., 11:4135–4151, (2011).

33. Liao TW. Two hybrid differential evolution algorithms for engineering design optimization. Appl. Soft Comput., 10:1188–1199, (2010).

34. Yi HZ, Duan Q, Liao TW. Three improved hybrid metaheuristic algorithms for engineering design optimization. Appl. Soft Comput., 13:2433–2444, (2013).

35. Mladenovic N, Brimberg J, Hansen P, Moreno-Perez JA. The p-median problem: A survey of metaheuristic approaches. Eur. J. Operat. Res., 179:927–939, (2007).

36. Liao TW, Egbelu PJ, Sarker BR, Leu SS. Metaheuristics for project and construction management - a state-of-the-art review. Autom. Constr., 20:491–505, (2011).

37. Zavala GR, Nebro AJ, Luna F, Coello CAC. A survey of multi-objective metaheuristics applied to structural optimization. Struct. Multidiscip. Optim., 49:537–558, (2014).

38. Blum C, Roli A. Metaheuristics in combinatorial optimization: Overview and conceptual comparison. ACM Comput. Survey, 35:268–308, (2003).

39. Alqahtani A. Kernel Ridge regression for prediction of streamflow in arid regions: a case study in Saudi Arabia. J. Hydrol., 563:199–212, (2018).

40. Hassan HM, Abdelrahman OM, Morsy MS. Development of a support vector machine model for predicting the elastic modulus of self-compacting concrete. Constr. Build. Mater., 249:118811, (2020).

41. Haghgoo M, Saadatzi MN. Neural network approach for predicting the ductility of steel sheets in deep drawing process. J. Mater. Eng. Perform., 28:123–130, (2019).

42. Jiang W, Li J, Chen C, Wang D, Li H. A machine learning approach to predicting thermal conductivity of polymers based on molecular dynamics simulations. Materials, 12:2088, (2019).

43. Ilyas IF, Rekatsinas T. Machine learning and data cleaning: Which serves the other? ACM J. Data Inform. Quality, 14:13, (2023).

44. Li W, Li W, Liu L, Li Y. Review of machine learning in structural health monitoring. Eng. Struct., 198:109481, (2019).

45. Mandal S, Adhikari S, Mahesh PA. Machine learning approaches in structural engineering: A review. J. Civil Struct. Health Monit., 11:777–803, (2021).

46. Ghosh S, Das S. Machine learning for civil infrastructure monitoring: a review. Struct. Infrastruct. Eng., 17:628–648, (2021).

47. Shorten C, Khoshgoftaar TM. A survey on image data augmentation for deep learning. J. Big Data, 6:60, (2019).

48. Sajjad M, Khan S, Muhammad K, Wu WQ, Ullah A, Baik SW. Multi-grade brain tumor classification using deep CNN with extensive data augmentation. J. Comput. Sci., 30:174–182, (2019).

49. Lemley J, Bazrafkan S, Corcoran P. Smart augmentation learning an optimal data augmentation strategy. IEEE Access, 5:5858–5869, (2017).

50. Alkadri AM, Elkorany A, Ahmed C. Enhancing detection of Arabic social spam using data augmentation and machine learning. Appl. Sci.-BASEL, 12:11388, (2022).

51. Chlap P, Min H, Vandenberg N, Dowling J, Holloway L, Haworth A. A review of medical image data augmentation techniques for deep learning applications. J. Med. Imag. Radiat. Oncol., 65:545–563, (2021).

52. Virkkunen, I, Koskinen, T, Jessen-Juhler, O, Rinta-aho, J. Augmented ultrasonic data for machine learning. J. Nondestruct. Eval., 40:4, (2021).

53. Seo J, Kapania RK. Topology optimization with advanced CNN using mapped physics-based data. Struct. Multidiscip. Optim., 66:21, (2023).

54. Gibson J, Hire A, Hennig RG. Data-augmentation for graph neural network learning of the relaxed energies of unrelaxed structures. NPJ Computat. Mater., 8:211, (2022).

55. Byun S, Yu JY, Cheon S, Lee SH, Park SH, Lee TKY. Enhanced prediction of anisotropic deformation behavior using machine learning with data augmentation. J. Magnes. Alloy., 12:186–196, (2024).

56. Wang H, Hu Y. Artificial intelligence and machine learning in civil engineering: A comprehensive review. Adv. Eng. Softw., 164:103912, (2021).

57. Lin T-S, Coley CW, Mochigase H, Beech HK, Wang W, Wang Z, Woods E, Craig SL, Johnson JA, Kalow JA, Jensen KF, Olsen BD. BigSMILES: A structurally based line notation for describing macromolecules. ACS Centr. Sci., 5:1523–1531, (2019).

58. Fan J, Li G. High enthalpy storage thermoset network with giant stress and energy output in rubbery state. Nat. Commun., 9:642, (2018).

59. Feng X, Fan J, Li A, Li G. Multi-reusable thermoset with anomalous flame triggered shape memory effect. ACS Appl. Mater. Interf., 11:16075–16086, (2019).

60. Feng X, Fan J, Li A, Li G. Biobased tannic acid crosslinked epoxy thermosets with hierarchical molecular structure and tunable properties: Damping, shape memory and recyclability. ACS Sustain. Chem. Eng., 8:874–883, (2020).

61. Feng X, Li G. Versatile phosphate diester based flame retardant vitrimers via catalyst-free mixed transesterification. ACS Appl. Mater. Interf., 12:57486–57496, (2020).

62. Feng X, Li G. High-temperature shape memory photopolymer with intrinsic flame retardancy and record-high recovery stress. Appl. Mater. Today, 23:101056, (2021).

63. Feng X, Li G. Catalyst-free β-hydroxy phosphate ester exchange for robust fireproof vitrimers. Chem. Eng. J., 417:129132, (2021).

64. Feng X, Li G. Room-temperature self-healable and mechanically robust thermoset polymer for healing delamination and recycling carbon fiber. ACS Appl. Mater. Interf., 13:53099–53110, (2021).

65. Feng X, Li G. UV curable, flame retardant, and pressure-sensitive adhesives with two-way shape memory effect. Polymer, 249:124835, (2022).

66. Wick C, Peters A, Li G. Simulation study of shape memory polymer networks formed by free radical polymerization. Polymer, 281:126114, (2023).

67. Nourian P, Wick C, Li G, Peters A. Correlation between cyclic topology and shape memory properties of an amine-based thermoset shape memory polymer: a coarse-grained molecular dynamics study. Smart Mater. Struct., 31:105014, (2022).

68. Shafe A, Wick C, Peters A, Liu X, Li G. Effect of atomistic fingerprints on thermomechanical properties of epoxy-diamine thermoset shape memory polymers. Polymer, 242:124577, (2022).

69. Lu L, Pan J, Li G. Recyclable high performance epoxy based on transesterification reaction. J. Mater. Chem. A, 5:21505–21513, (2017).
70. Li A, Fan J, Li G. Recyclable thermoset shape memory polymer with high stress and energy output via facile UV-curing. J. Mater. Chem. A, 6:11479–11487, (2018).
71. Lu L, Cao J, Li G. Giant reversible elongation upon cooling and contraction upon heating for a crosslinked cis poly(1,4-butadiene) system at temperatures below zero Celsius. Sci. Rep., 8:14233, (2018).
72. Feng X, Li G. Photo-crosslinkable and ultrastable poly(1,4-butadiene) based organogel with record-high reversible elongation upon cooling and contraction upon heating. Polymer, 262:125477, (2022).
73. Sarrafan S, Li G. Conductive and ferromagnetic syntactic foam with shape memory and self-healing/recycling capabilities. Adv. Funct. Mater., 34:2308085, (2024).
74. Challapalli A, Li G. 3D printable biomimetic rod with superior buckling resistance designed by machine learning. Sci. Rep., 10:20716, (2020).
75. Wang A, Li G. Stress memory of a thermoset shape memory polymer. J. Appl. Polym. Sci., 132:42112, (2015).
76. Challapalli A, Patel D, Li G. Inverse machine learning framework for optimizing lightweight metamaterials. Mater. Des., 208:109937, (2021).

4 Structural Optimization of Biomimetic Rods Using Machine Learning Regression

4.1 INTRODUCTION

Biomimicry, the emulation of natural processes, components, and architectures in engineering, draws inspiration from the intricate designs found in nature [1]. Humans have long been inspired by biological systems. By observing the bird fly, humans started to build the first aircraft. Actually, almost all human designs can find a counterpart in nature. For example, gecko setae were mimicked to produce dry adhesive [2–5]. Woodpecker was simulated to design impact tolerance structures [6–10]. Nacre was mimicked to build strong and yet tough materials [11–13]. Even today, with powerful computation and simulation tools, many designs and inventions are inspired by nature's counterparts. With millions and millions of years of evolution, natural structures such as biological systems have developed amazing structures with superior load-carrying capacities and multifunctionalities, by using a small number of simple building blocks such as proteins and ceramics. Natural structures, honed by millions of years of evolution, exhibit remarkable efficiency and resilience [14]. Through the process of natural selection, organisms have developed sophisticated strategies to adapt to their environments and optimize their performance. For example, the hierarchical organization of materials in biological systems, such as the arrangement of collagen fibers in bone, inspired the development of bioinspired materials with enhanced mechanical properties [15]. By observing and understanding biological systems, researchers seek to develop innovative solutions to complex engineering challenges. The concept of biomimicry has gained significant attention across various fields, including materials, architecture, biomedical devices, robotics, structures, manufacturing, and engineering [16–39].

Biomimicry has found widespread applications in engineering, leading to the development of novel materials, biomimetic robots, and sustainable architectures [40]. By mimicking the structural features and functional principles of biological systems, engineers aim to overcome design limitations and achieve superior performance in engineered products. For instance, the design of lightweight structural components inspired by natural materials, such as bamboo and bird bones,

DOI: 10.1201/9781003400165-4

has led to the development of high-strength, low-density materials for aerospace and automotive applications [41].

In architectural design, biomimicry offers a focus towards sustainable and resilient built environments [42]. By studying the efficient use of materials and energy in natural systems, architects and designers can create buildings that harmonize with their surroundings and minimize environmental impact. For example, the design of biomimetic facades inspired by the thermoregulatory mechanisms of termite mounds has led to buildings with enhanced energy efficiency and indoor comfort [42].

The study of natural structures, such as plant stems and animal skeletons, provides valuable insights for optimizing the performance of structural elements in engineering applications [41]. For instance, the porous architecture of plant stems, which enhances their resistance to bending and buckling, has inspired the design of biomimetic columns with improved load-bearing capacity [42]. Similarly, the hierarchical structure of marine shells, characterized by intricate patterns of interconnected pores, has inspired the development of lightweight and impact-resistant materials for aerospace and defense applications [43].

The integration of machine learning techniques with biomimicry offers new avenues for the design and optimization of engineered systems [44]. Machine learning algorithms can analyze vast datasets of biological structures and identify patterns that inform the design of biomimetic materials and components [45]. For example, researchers have used machine learning algorithms to analyze the relationship between the microstructure of biological materials and their mechanical properties, leading to the development of bioinspired materials with tailored properties [46].

The field of biomimicry continues to evolve, with ongoing research focused on expanding our understanding of natural systems and translating this knowledge into innovative engineering solutions [47]. Future developments may include the integration of advanced manufacturing techniques, such as 3D printing and nanotechnology, to fabricate biomimetic structures with unprecedented precision and complexity [48]. Additionally, interdisciplinary collaborations between biologists, engineers, and computer scientists will be essential for driving forward the field of biomimicry and unlocking its full potential for addressing global challenges [49–53].

In load-bearing structures, columns, rods, or cylinders are a fundamental type of structural element. It is well known from fundamental structural mechanics that, under axial compression, these columns may suddenly lose stability, i.e., bend laterally, at loads well below the compressive strength of the materials used to manufacture the columns. This phenomenon has been well documented as buckling, and the Euler buckling equation is widely used to estimate the buckling load. In the design and optimization of load-bearing structures, buckling must be avoided. From the buckling load equation, it is known that buckling is related to the bending stiffness, in addition to the length and boundary conditions of the column. Clearly, the materials around the midplane of the column do not contribute very much to bending stiffness. In other words, solid columns do not have an advantage over hollow columns in terms of buckling resistance. This is why T-beam, box-beam, and H-beam are widely used in engineering structures, instead of solid beams. Therefore, porous columns are highly desired. In an attempt to develop more resilient rods, researchers have turned to the natural world for inspiration. The objective of the research is to create columns or rods that exhibit superior resistance to buckling by drawing on the features and characteristics of biological structures. From the above discussion, porous configurations are not only prevalent in plants but also exist in marine shells, animal quills, and honeycombs, as indicated in the diagram on the left side of Figure 4.1. On the right side of Figure 4.1, these biological structures can be represented as idealized biomimetic structures. Different biomimetic rods represented in Figure 4.1 are summarized in Table 4.1. Due to our comprehension of buckling resilience, these porous structures can be considered as excellent alternatives for creating biomimetic rods that possess superior resistance against buckling.

To accomplish this task, researchers first create a digital representation of each biological structure, recording its unique features and properties to create a database. Later machine learning algorithm is used to analyze the database, identifying correlations between the biological features and buckling load. Finally, they filter the data to optimize the biological structures, creating idealized models that are modeled using finite element analysis (FEA) to obtain their buckling loads.

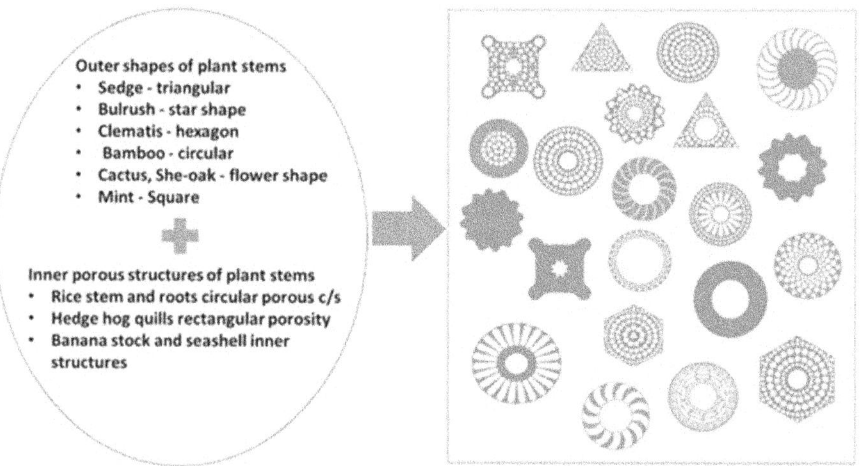

FIGURE 4.1 Biomimetic inspirations and depiction of 2D cross-section. The figure contains different outer shapes and inspirations such as sedge inspired for triangular shape, bulrush plant stem which has a star shape, clematis inspiration for hexagonal shape, bamboo inspiration for hallow shape, cactus, and she-oak stems inspiration for flower shapes, mint stems with a square like shapes combined with inner shapes inspiration based on rice stem, and hedgehog stock and sea shell inner structures to create different porous structures. On the righthand side of the image, several combinations of these outer and inner shapes are depicted, which form the training dataset for biomimetic columns.

TABLE 4.1

Different Biomimetic Rods Represented in Figure 4.1

t1	Hedgehog quill cross-section 
t2	Cactus stem dummy rod 
t3	Square stem 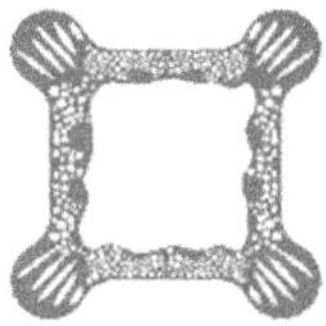
t4	Banana stem 
t5	She-oak stem 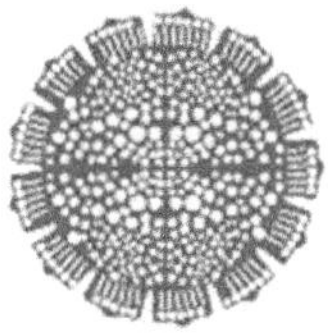
t6	Sedge stem 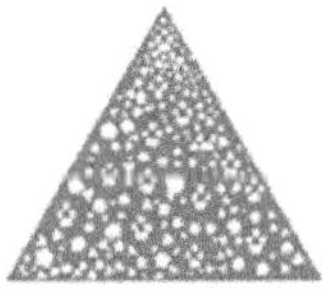
t7	Square stem dummy rod 

(Continued)

TABLE 4.1 *(Continued)*
Different Biomimetic Rods Represented in Figure 4.1

t8	Clematis (Hexagonal) stem	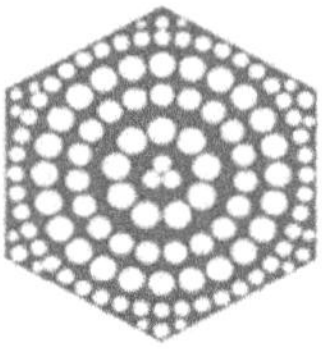
t9	Nymphaea stem	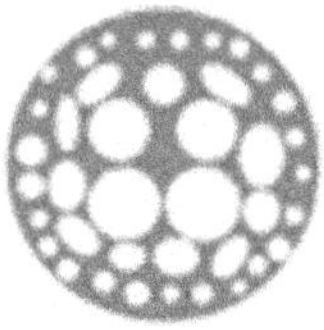
t10	Seashell	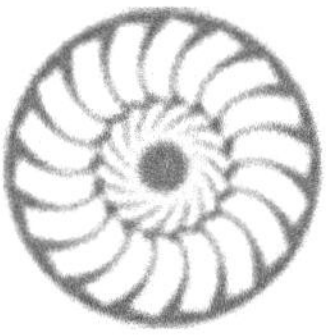
t11	Hollow cylinder with groves (Inspired from seashells)	
t12	Bamboo stem	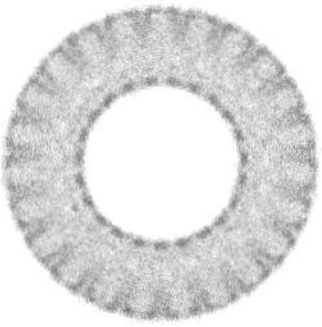
t13	Bamboo stem dummy rod	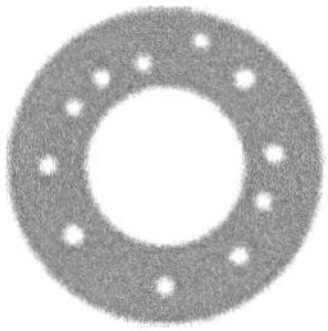

Validation of the FEA predictions is then carried out by 3D printing and testing some typical biomimetic rods. Unlike solid rods under ideal boundary conditions, such as simply supported or fixed boundary conditions, the complex geometrical shape of biomimetic rods makes the calculation of bending stiffness extremely difficult, which limits the direct use of the Euler buckling equation. Therefore, FEA is employed to accelerate the discovery of optimized biomimetic rods as seen in Figure 4.2.

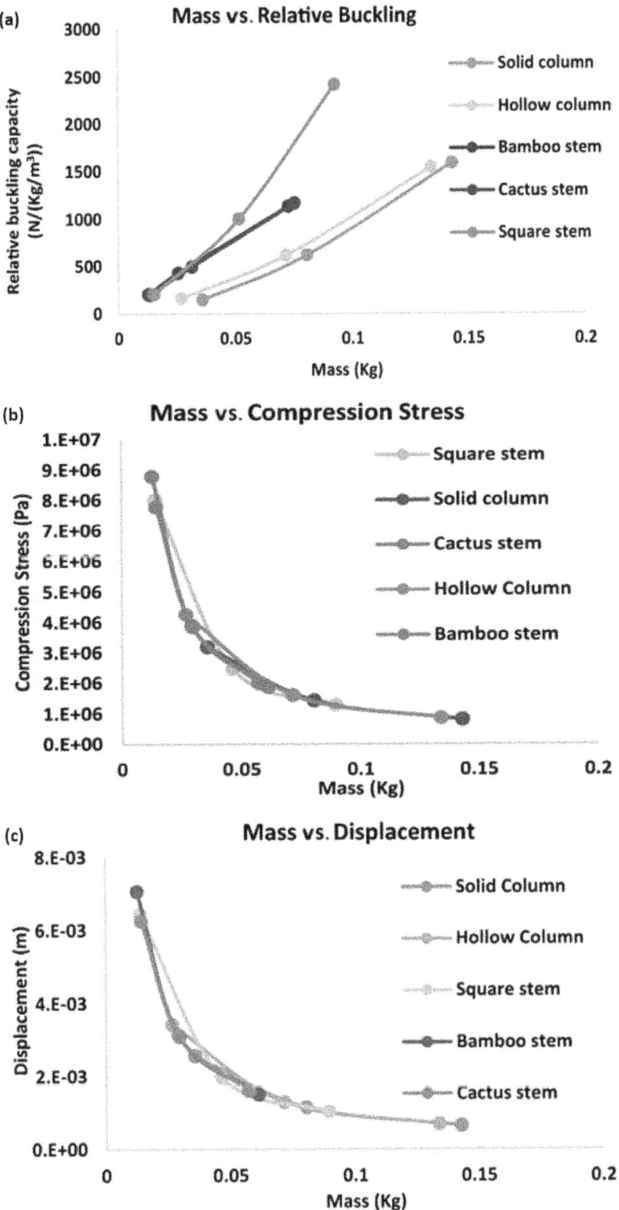

FIGURE 4.2 The outcomes of ANSYS simulations were examined for control solid rods, hollow rods, and biomimetic rods. The simulation results were represented in terms of (a) normalized buckling capacity versus mass comparing different biomimetic rods with solid and hollow rods, (b) compression stress, and (c) axial displacement. The results show that there is improvement in the buckling capacity in the biomimetic rods, while the compression stress and axial displacements of the biomimetic rods are on par with solid and hollow cylinders.

It is worth noting that the research aims not only at mimicking nature but also at improving upon it. In other words, the objective of biomimetic design is to learn from nature but better than nature. For rods under axial loading, the result is that biomimetic rods surpass their natural counterparts in their resistance to buckling. Although through biomimicry, optimal designs were able to be generated, it is believed that there exists a large design space that even the biological structures have not occupied. For example, by changing the shapes or by having various combinations of inner and outer shapes of the bio-mimetic designs proposed above in Figure 4.1, thousands of new designs can be created. In order to enhance the performance of biomimetic rods, the next step is to optimize their design to achieve even greater buckling strength. One approach for optimization is to use machine learning algorithms, which offer significant advantages in terms of speed and ease of design. To use machine learning for optimization, the algorithm needs to be trained on a set of finger-prints of known rod designs and then used to predict the buckling loads of new, untrained designs. In this chapter, the following sections will go in-depth into the steps followed to generate, fingerprint, train, and evaluate the data, and predict new biomimetic rods.

4.2 SELECTION OF BIOLOGICAL COUNTERPARTS AND CREATION OF BIOMIMETIC RODS

In this study, biomimetic rods with a combination of the external shape (upper left in Figure 4.1) and internal microstructure (bottom left in Figure 4.1) in biological counterparts are created as shown on the right-hand side in Figure 4.1. From the schematics on the right-hand side in Figure 4.1, we created 21 basic biomimetic rods. In particular, the external stem shapes of the rice plant, bamboo, cactus, square (mint, cup plant), bulrush, papyrus, and she-oak plants, and the internal porous structures of roots, hedgehog quills, seashells, and honeycomb were com-bined to create the biomimetic rods. However, to conduct machine learning, 21 rods are not enough. In this study, a total of 1,500 rods were created for model-ing. These additional rods were created based on the 21 basic rods, by making modifications such as changing the shape of pores, pore size distributions, pore locations in the rods, etc. The number of additional rods created for each group can be found in Table 4.2.

In Table 4.2, the biomimetic rods were arranged into 7 groups based on their external shape. To keep consistency, each column or rod had the same height of 10 cm and the same overall volume (volume of the solid material + volume of the pores). We used a circular rod with a diameter of 1 cm to determine the volume, which was 7.85 cm^3. For comparison purposes, both solid cylinder and hollow cylinder (1 cm outer diameter and 0.5 cm inner diameter) were used as control rods. In addition to the 1 cm outer diameter for both solid and hollow rods, two additional outer diameters of 1.5 cm and 2.0 cm were also investigated, leading to a total of 6 control rods.

TABLE 4.2

The Created Biomimetic Rods for Finite Element Analysis and Training Dataset

Biomimetic Inspiration for External Shape	Primary Designs for Internal Microstructure (a Total of 21)	Extended Dataset by Adjusting the Internal Structures (a Total of 1,500)
Bamboo, rice		91 dummy rods
Square plant		81 dummy rods
Bulrush		71 dummy rods
Cactus		81 dummy rods
She-oak		50 dummy rods
Sedge		50 dummy rods

To generate these new designs, a MATLAB code was developed to create all possible combinations of structural features inspired by the biomimetic rods. This process generated over a million combinations, but not all of them exhibited superior structural properties. Through manual inspection of a few initial datasets, a clear pattern emerged which helped to establish a range of porosity values that produced better-performing rods compared to the semi-optimal biomimetic rods. The MATLAB code is listed below:

```matlab
% Define the vector
        v = [1 0 0; 9 0 0; 14 0.8 -0.18333; 14 0.71667
-0.26667; 14 0.56667 -0.31667; 14 0.76667 -0.51667; 14
0.56667 -0.51667; 14 0.36667 -0.53333].
% Get the length of the vector
        n = length(v);
% Initialize the output cell array
        combinations = cell(1, 2^n - 1);
% Loop through all possible combinations
        for i = 1:2^n - 1
% Convert the index to binary and pad with zeros
        binary = dec2bin(i, n);
        binary = ['0' binary];
% Initialize the current combination
        current = [];
% Loop through each element of the vector
        for j = 1:n
% Check if the jth bit is 1
        if binary(j) == '1'
% Append the jth element to the current combination
            current = [current; v(j,:)];
        end
      end
  % Add the current combination to the output cell array
      combinations{i} = current;
end.
```

This code first defines the input vector v and then gets its length n. It then initializes an output cell array combination with a length of 2^n (since there are $2^n - 1$ possible non-empty combinations of a vector of length n). The code then loops through all possible combinations by looping through all numbers from 1 to $2^n - 1$, converting each index to binary, and then looping through each element of the vector and checking whether the corresponding bit is 1. If the bit is 1, the jth element of the vector is added to the current combination. Finally, the current combination is added to the output cell array. At the end of the code, the combinations cell array contains all possible non-empty combinations of the input vector.

4.3 BUCKLING LOAD ANALYSIS OF THE BIOMIMETIC RODS

In order to create a dataset for machine learning, the buckling load of the biomimetic rods in Table 4.2 must be known. However, for the biomimetic rods created in Table 4.2, their buckling load is largely unknown. Therefore, finite element analysis (FEA) was used to determine the buckling load for each biomimetic rod. ANSYS workbench was used to conduct stress and buckling analysis. To provide the constitutive equation for the FEA, cylindrical samples of a chosen polymer were 3D-printed. In this study, 3D printable Polylactic Acid (PLA) manufactured by Hatchbox was used as the common material for all the designs. The mechanical properties under uniaxial compression of PLA were evaluated using a Q-TEST 150 machine. For this purpose, ASTM standard D695-15 for compression of additive manufactured components was followed. Ten samples were printed with five for the strength test and the other five for the modulus test. The speed of testing, which was 2mm/min., and measurement procedures followed the ASTM D695-15 specifications. The properties of the material are summarized in Table 4.3. The stress vs. strain data from the uniaxial compression test at room temperature was imported directly into ANSYS as the constitutive law; see a typical result in Figure 4.3.

TABLE 4.3

Mechanical Properties of 3D Printable PLA

Material	PLA
Density	1,138 kg/m^3
Poisson's ratio	0.4
Young's modulus	1.24 GPa
Compressive strength	55.30 MPa
Tensile strength	11.00 MPa

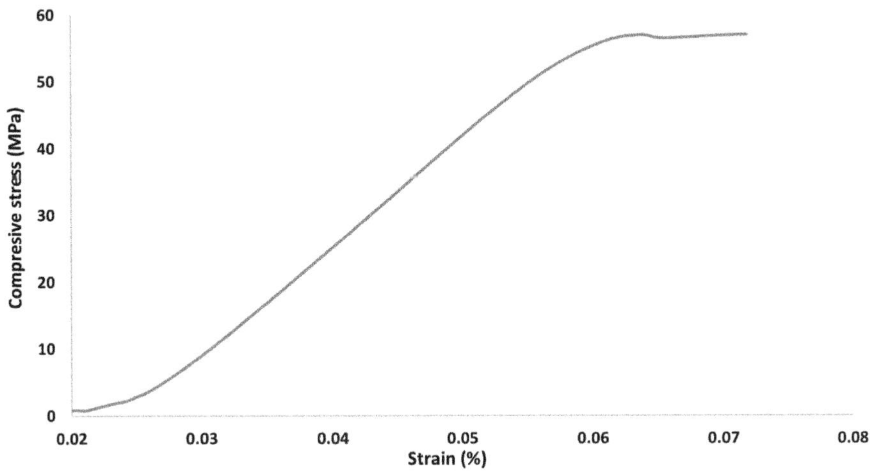

FIGURE 4.3 Stress vs strain curve for a typical 3D-printed compression test sample.

All the rods in Table 4.2 were modeled using ANSYS design modeler and analyzed under uniaxial compression. The same height (10 cm), boundary conditions, loading and meshing scheme were applied to all the rods for consistency. Static analysis and Euler's buckling analysis were conducted to evaluate the rod response at a constant load. For buckling analysis, all the rods were modeled with the same uniaxial compression load (1,000 N) with one end fixed and the other end pin supported. For each rod, ANSYS outputs a buckling factor, which when multiplied by the applied load (1,000 N), gives the actual buckling load of the rod. A convergence analysis was conducted, and based on the results, the final element type was determined to be hexahedral, and the element size was 0.1 mm. The buckling load, stress, displacement, mass, and volume of all the rods were recorded from the model interface.

4.4 EXPERIMENTAL VALIDATION

To make sure that the FEA can capture the buckling behavior of the biomimetic rods, experiments were conducted, and the results were compared with the FEA results. For the experimental validation, initially SolidWorks design tool was used to model some typical biomimetic rods. STL file formats were generated for the rods in order to be machine-readable. An extrusion-based 3D printer-Creality Cr-10s at Louisiana State University (LSU) was used to print these biomimetic rods using PLA filament. PLA with an extruder temperature of 210°C and nozzle thickness of 1.75 mm was used. The 3D printer consists of three different resolutions from coarse to fine layer thickness. To compensate for the time constraint and maintain consistency, coarse resolution was employed for all the rods printed. Minimum post-processing was required for extrusion-based printing of these designs as only a few supports were required. The STL files of the drawings from SolidWorks were converted into a g-code using the Cura [54] interface available online. These g-codes were fed into the printer to print the desired rods. Figure 4.4 shows several 3D-printed biomimetic rods. Once all the rods were printed and cleared of the support material, a Q-TEST 150 testing machine with a capacity of 150 KN was used for the compression testing. The test was conducted at 2 mm/min for all the rods. The mass of each rod was physically weighed before the tests. The buckling capacities, i.e., the load when the rods start to buckle were recorded.

Comparisons between the 3D-printed biomimetic rods and FEA-modeled rods in terms of normalized buckling loads are depicted in Figure 4.5. It is seen that the experimental results agree with the simulation results. A slight difference is found between the experimental and simulation, which is primarily due to the compromise in the 3D printing due to low printing resolution [55]. It should be noted that the boundary conditions for the rods in the simulations were adjusted to suit the experimental conditions. Table 4.4 gives the comparison of the buckled shape between the FEA modeling and experimental testing for several typical biomimetic rods, again, with a good agreement. Therefore, the FEA is validated. The validated FEA will be utilized to calculate the buckling load of the biomimetic rods in the training database later.

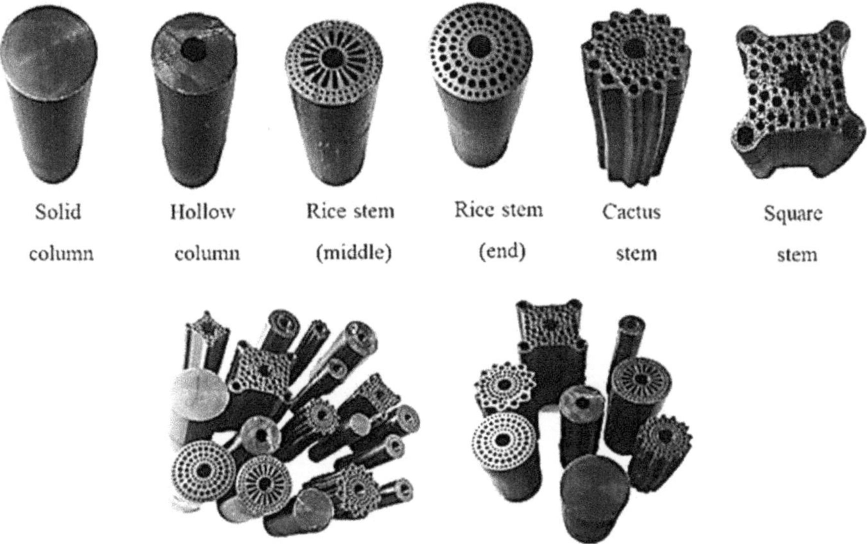

FIGURE 4.4   3D printed biomimetic rods.

From this initial investigation, it was observed that directly mimicking these structures in obtaining optimum designs has a couple of limitations. Firstly, it is too strenuous and not ideal to manually design, analyze, and compare all the possible rods. Secondly, although an optimum set of structures can be deduced from the designs, it is believed that much better designs with higher structural

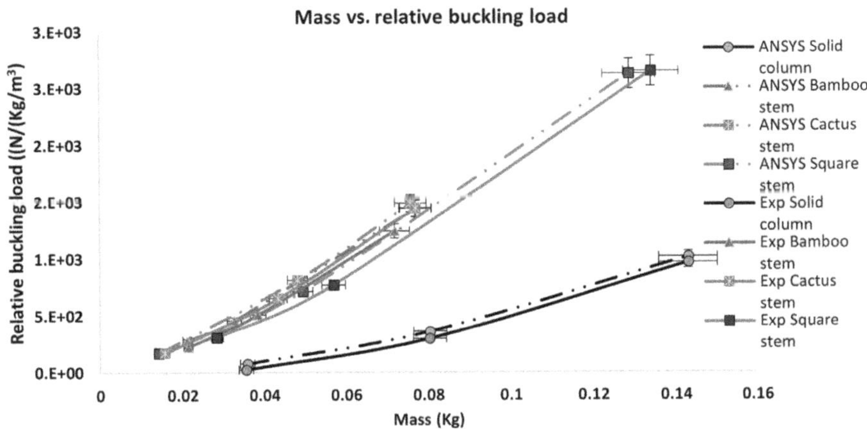

FIGURE 4.5   Experimental validation for buckling capacity conducted on biomimetic rods. Four groups of biomimetic rods were used for validation: bamboo stem, cactus square, square stem, and solid.

TABLE 4.4

Buckling Mode Comparisons of Different Rods (A) FEA and (B) Experiments

Name	(A) Deformed Shape by ANSYS	(B) Deformed Shape by Experiment
Solid column		
Bamboo stem		
Cactus stem		

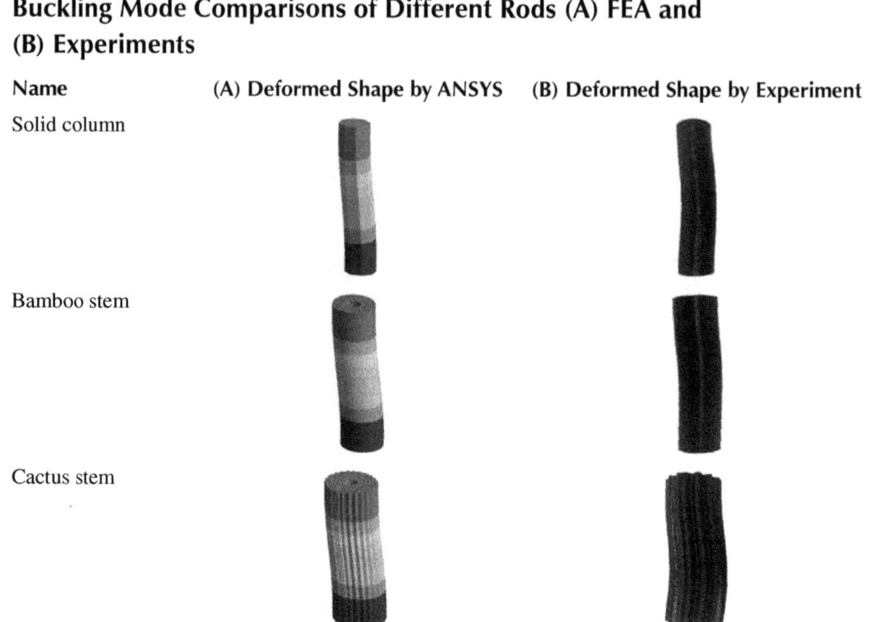

capacities exist and it is impossible to explore all the possible structures or the structure design space by combining these biomimetic designs. Therefore, machine learning was used to help identify the optimal rods.

4.5　FEATURE IDENTIFICATION OR FINGERPRINTS

The first step in machine learning is to digitalize the proposed biomimetic rods so that a training dataset can be obtained. Machine learning is artificial intelligence which can be used to train systems to learn from the data provided, which in turn can be used for predicting or categorizing new untrained data. It is very advantageous in reducing human intervention, complicated programming, and computational time. In order to conduct machine learning, the biomimetic rods must be converted to computer recognizable format, or feature identification or fingerprinting. In other words, the machine learning algorithm needs the rods to be fingerprinted. Fingerprinting refers to converting each rod into a machine-readable code or sequence. For this purpose, each different shape and feature of the designs is given a unique identification, a scalar or a vector, or a matrix. For example, a bamboo-inspired column consists of an outer circular shape, an inner circular surface, and several smaller hollow cylinders of different diameters. These shapes differ from design to design. A unique number for the outer circle, inner circle, and the rest of the smaller circles are assigned along with their location with respect to the origin of a coordinate system. It was made sure that the outer circle's center lay on the origin of the coordinate system for all the designs, i.e., all the substructures share the same global coordinate system. For simplification, all the small circles that form porosity are designed to be of the same diameter. Similarly, each design with a new shape is given a unique number. Seven outer circles (1 to 7), six inner circles (8 to 13), and three smaller inner circles (13 to 15) are identified from the bamboo-inspired designs. These shapes are numbered and used to form the fingerprints. A similar procedure is followed for all the 1,500 biomimetic rods in Table 4.2. A single vector consisting of different numbers including all the features in a rod was the final fingerprint for the rod. Some designs such as the root cross-section contain about 400 small spherical pores, which implies that the fingerprint vector comprises 400 variables plus the outer circle. For example, the fingerprint of a rod with outer, inner, and ten small cylinders is of the form "1 (0,0), 9 (0,0), 14 (0.8, −0.18333), 14 (0.71667, −0.26667), 14 (0.56667, −0.31667), 14 (0.76667, −0.51667), 14 (0.56667, −0.51667), 14 (0.36667, −0.53333), 14 (0.45, −0.71667), 14 (0.31667, −0.65), 14 (0.11667, −0.6), 14 (0.21667, −0.75)." It uniquely defines a rod that contains an outer circle (1 (0,0)), an inner circle (9 (0,0)) and ten smaller circles (14 (0.8, −0.18333), 14 (0.71667, −0.26667), 14 (0.56667, −0.31667), 14 (0.76667, −0.51667), 14 (0.56667, −0.51667), 14 (0.36667, −0.53333), 14 (0.45, −0.71667), 14 (0.31667, −0.65), 14 (0.11667, −0.6), 14 (0.21667, −0.75)). The meaning of the numbers can be explained as follows. For example, for 14 (0.21667, −0.75), the number 14 means it is a small circle or cylinder, and (0.21667, −0.75) is the coordinate of the bottom center of the cylinder.

4.6 FORWARD DESIGN AND PREDICTION

Machine learning discovery of biomimetic rods can be divided into two approaches: forward machine learning and inverse machine learning. In forward machine learning, the machine learning algorithm establishes the correlation between the fingerprinting (inputs) of the rods and the buckling load (output). Mathematically, this approach has many independent variables but only one or few dependent variables and is comparatively easier to solve. Once the machine learning model is validated through numerical analysis such as FEA, or experimental testing, or both, the model can be used to discover new rods. In this process, many new fingerprints can be created, and they can be input to the machine learning model, and each input will lead to a new output, which is buckling load. A comparison of the output buckling load with the threshold buckling load can serve as a screening approach. The biomimetic rods with buckling load higher than that of the threshold value can be selected as candidates for further evaluation and may become the newly discovered biomimetic rods.

Another approach is inverse design. In this approach, the machine learning model establishes the correlation between the desired buckling load and fingerprints of the biomimetic rods. Mathematically, this approach is more challenging because it has one or a small number of independent variables but a huge number of dependent variables. In this study, we use forward design to discover new biomimetic rods with superior buckling load.

Machine learning can be formally perceived in two settings, supervised and unsupervised learning. Supervised learning implies data having known input and output values. On the contrary un-supervised learning consists of a set of inputs without labels. In this study, the data available (biomimetic rods with uniaxial buckling loads) have both the inputs, which are the structural properties such as mass, volume and microstructure, and the output, which is the buckling strength of individual structures (rods). Hence, supervised learning was used to train the dataset. All the rods were fingerprinted in the same manner as proposed above to maintain consistency. The mass, volume, and buckling strength of all the designs were used directly as features. All the data were stored in an Excel sheet containing the mass, volume, buckling strength, and fingerprints of geometrical features of the individual rods. While approaching a supervised or unsupervised learning problem, many different algorithms are available and no one approach can be called superior to the others. Each type of problem and datasets have a suitable algorithm that works best. Supervised machine learning algorithms are being widely used in material and chemical engineering for property prediction and discovering new materials. Kernel Ridge Regression (KRR) has been used to handle non-linear relations in the property predictions of polymers [56]. Gaussian Process Regression (GPR) has been used by the same group identifying that it is more suitable for predicting a better uncertain/confidence interval of polymers and their properties [56]. Neural network models have been used by Stephen et al. to assist in the discovery of polymers with high thermal conductivity [57]. Support vector machines (SVM) which are considered to be very effective in real value function estimation are used to predict the mechanical properties of cement [58, 59]. Several other algorithms like decision trees, K-Nearest Neighbors, and gradient boosting algorithms have been proven to be effective in predicting properties with high accuracy [60]. Neural networks are used to estimate the stress distribution in the aortic wall based on FEA results with an average discrepancy of 0.492% [61]. A regression tree is used to predict the mechanical properties of carbon fibers like the longitudinal and transverse elastic modulus and shear modulus using data generated from finite element modeling [62]. Support Vector Regression models are used to propose FEA models which can find a direct relationship between the input and output of the elements. This avoids the complex numerical iterations involved in finding the internal displacement field [63]. Ensemble methods were used to model the biomechanical behavior of breast tissues under compression using results for FEA models [64].

In this study, MATLAB was used to evaluate the dataset with different machine learning algorithms available. Ninety percent of the data was used for training the regression models and the remaining ten percent of the data was used for testing the performance of the various algorithms. The mass, volume, and fingerprints were defined as the inputs, and the buckling strength was defined as the output for all the regression models. Since the output is a single variable depending on multiple input variables that are correlated to each other, it is easy for the machine learning algorithm to formalize a relationship between them. Five-fold cross-validation is used to evaluate the performance of the machine learning algorithms in predicting new data. MATLAB allows us to check the performance of various pre-programmed machine learning algorithms with our data and gives a root mean square error (RMSE) of each model prediction. It is the difference between the predictions and actual values of the observations. It was observed that the Ensemble Bagged Tree algorithm with a leaf size of eight is the best-suited machine learning algorithm based on the RMSE as compared to other models like SVM, GPR, and neural networks for the data type being analyzed. In this model, an Ensemble tree divides the dataset into different subsets and trains each subset individually. The average of all the subset model predictions was used as the final prediction [65]. The Ensemble of all the subsets gave a much more robust model as compared to individual subsets.

Figure 4.6 shows the predictions by machine learning vs. true response or observations by FEA. With the line being predictions and the dots being the observations, the closer the observations to the prediction line, the better the model is. From Figure 4.6, it is clear that the ensemble tree model gives the best prediction.

In the ensemble tree approach, the data set is randomly divided into different subsets by the algorithms within itself for predictions. Though precise values for predictions may not be achieved by an ensemble tree as it is based on the mean of predictions from the subsets, this model is very advantageous in handling sophisticated data. Since the fingerprints of the current data type contain very large vectors, the ensemble tree seems to be a fitful model for regression with this particular data type. Once the model was ready it was exported to the MATLAB workspace for testing. The test data was made into a table matching the labels as the training data and imported into the workspace. The 'yfit' functions were used to predict the buckling strengths of the test data. The error percentage was calculated to be less than 10% for the majority of the test data, except for a few dummy points where it exceeded 10%. This margin of errors has also been used previously by others [56, 58, 59]. It is noted that, although the margin of errors has been used, a better comparison of the root mean squared error (RMSE) between the training and testing data would be useful when commenting on any overfitting, which can be an issue with ensemble trees. This will be considered in our

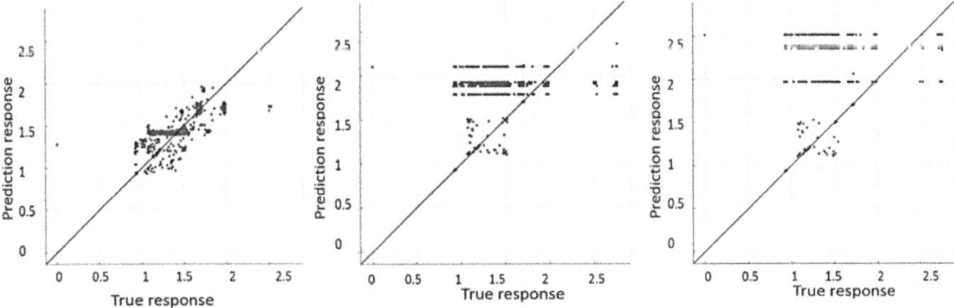

FIGURE 4.6 Comparison of predictions vs. true responses. From left to right, they are Ensemble tree, Support Vector machines, and Gaussian process regression for buckling capacities of the training dataset.

future studies. Hence this model can be used to predict the buckling strength of any column containing the features of the trained models instantly.

Sample code:

```
x_values = [1 (0,0), 2 (0,0), 3 (0,0), 4 (0,0), 5 (0,0), 6
(0,0), 7 (0,0), 1 (0,0), 8 (0,0), 1 (0,0), 11 (0,0), 1
(0,0), 13 (0,0), 1 (0,0), 9 (0,0), 1 (0,0), 12 (0,0), 1
(0,0), 10 (0,0), 4 (0,0), 8 (0,0), 4 (0,0), 11 (0,0), 4
(0,0), 13 (0,0), 4 (0,0), 9 (0,0), 4 (0,0), 12 (0,0), 4
(0,0), 10 (0,0)];
y_values = [1.6587, 1.9767, 1.7468, 1.9884, 2.4991, 1.7132,
1.6727, 1.5233, 1.5013, 1.52, 1.5386, 1.4803, 1.5129,
1.8148, 1.8003, 1.8111, 1.7688, 1.537, 1.839];
% Create a TreeBagger model for regression
NumTrees = 100; % Number of trees in the ensemble
Mdl = TreeBagger(NumTrees, x_values, y_values, 'Method',
'regression');
% Example prediction on new data
new_x = 8; % New x value
predicted_y = predict(Mdl, new_x);

disp(['Predicted y for new x: ', num2str(predicted_y)]);
```

Replace the placeholders x_values and y_values with your actual data. x_values should contain the predictor feature values (e.g., x coordinates) and y_values should contain the corresponding response variable values (e.g., y coordinates).

Create the Bagged Tree Model by specifying the number of trees in the ensemble using NumTrees.

Initialize the model using TreeBagger with the specified number of trees, predictor features (x_values), and response variable (y_values). The 'Method', and 'regression' argument indicates that a regression model is being created.

To predict the response for a new x value, use the predict function, and replace new_x with the desired value for which prediction is required to the corresponding y.

4.7   DISCOVERY OF NEW BIOMIMETIC RODS

Now we have validated our forward machine learning design model. We can then use the model to discover new biomimetic rods which have even higher buckling resistance than what we have in the training dataset, i.e., the 1500 biomimetic rods given in Table 4.2.

Therefore, the further step in designing biomimetic rods is to optimize these designs to discover even better rods with superior buckling resistance. The advantage of using a machine learning program for optimization is the exceptional speed in validating new designs and also ease in designing. The program needs to be fed with just the fingerprints of the new designs to predict their buckling loads. For optimization, the developed machine learning algorithm is used for forward prediction of various untrained fingerprints. It is fairly easy to develop a code that generates different fingerprint patterns compared to manually designing each structure individually. Therefore, initially a MATLAB code is programmed to generate all the possible combinations inspired by the biomimetic rods. This generates more than a million combinations. However, not all combinations exhibit finer structural properties. From the first few datasets, a clear pattern (number of internal microstructures) can be manually observed which helped in defining a minimum and maximum porosity for the rods to perform better compared to the semi-optimal (1,500 proposed biomimetic) rods in Table 4.2. This initially helped in truncating the data sets manually. For further filtering of non-optimal designs, EXCEL and MATLAB functions are (can be) used. In EXCEL, "IF" function and ">" or "<" can be used to identify the numbers greater or less than a given value (buckling load in this case) and the "Index" function can be used to display the filtered fingerprints. MATLAB does not have a function for indexing. But the ">" or "<" can be used to identify the desired fingerprints from the predicted dataset and the identified variables (fingerprints) can be called and defined into a new dataset. Finally, a dataset that contained 160 new designs (fingerprints) which exhibited better buckling properties compared to the biomimetic rods in the training dataset was generated. The optimum designs which have higher buckling strengths were taken to evaluate their performance. These new fingerprints were converted into 3D CAD designs using ANSYS and analyzed under uniaxial compression for their structural properties.

Figure 4.7 shows the machine learning framework, including potential applications in biomimetic lattice structures. After the optimization, a total of 160 new

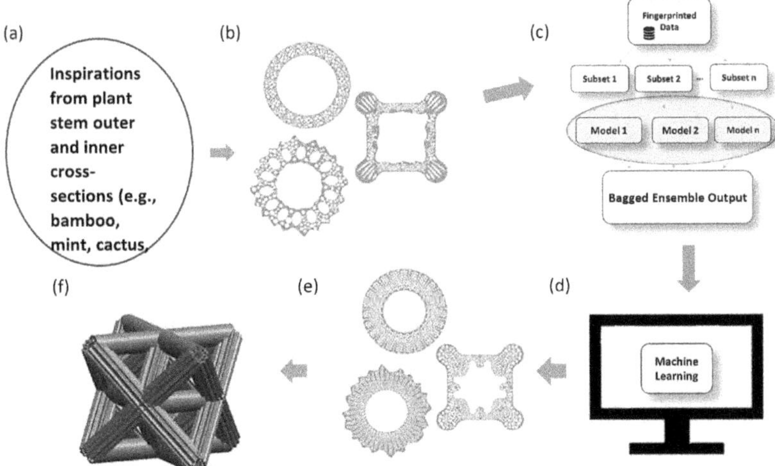

FIGURE 4.7 Machine learning assisted design of optimal biomimetic rods and lattice unit cell. (a) biomimetic inspirations, (b) initial designs, (c) forward machine learning predictions through bagged ensemble tree, (d) data filtering to find optimal biomimetic structures, (e) cAD designs of single optimum rods (top view), and (f) potential application (3D-printed lattice structure containing optimal bio-inspired rods).

TABLE 4.5

The Created Optimal Designs from Inverse Design and Optimization

External Shapes of Optimized Biomimetic Designs	Primary Designs with Optimized Internal Shapes	Number of Optimal Designs in Each Type
Bamboo		50 rods
Square plant		40 rods
Cactus		50 rods
Bulrush		20 rods

rods were created, which are all better than the 1,500 rods in the initial training database. Table 4.5 summarizes the 160 new designs.

In order to determine the buckling load of the newly discovered rods, the validated FEA was used again. Uniaxial compression modeling conducted on these new designs used the same boundary conditions as the modeling conducted on the newly discovered biomimetic rods. The results are reported in Figure 4.8. It is seen from Figure 4.8 that the optimized rods inspired from the biomimetic rods through optimization exhibit a buckling strength nearly double that of the biomimetic rods in the initial training dataset. Figure 4.9 shows the stress distribution for several optimized rods.

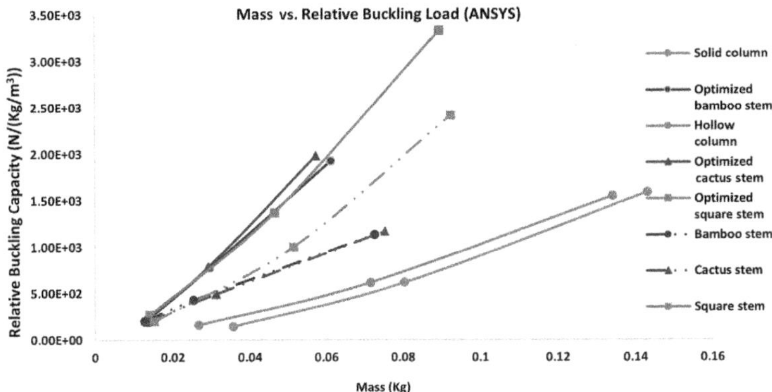

FIGURE 4.8 Simulation results from ANSYS conducted on various optimized rods through machine learning, as well as representative rods from the initial 1,500 training rods.

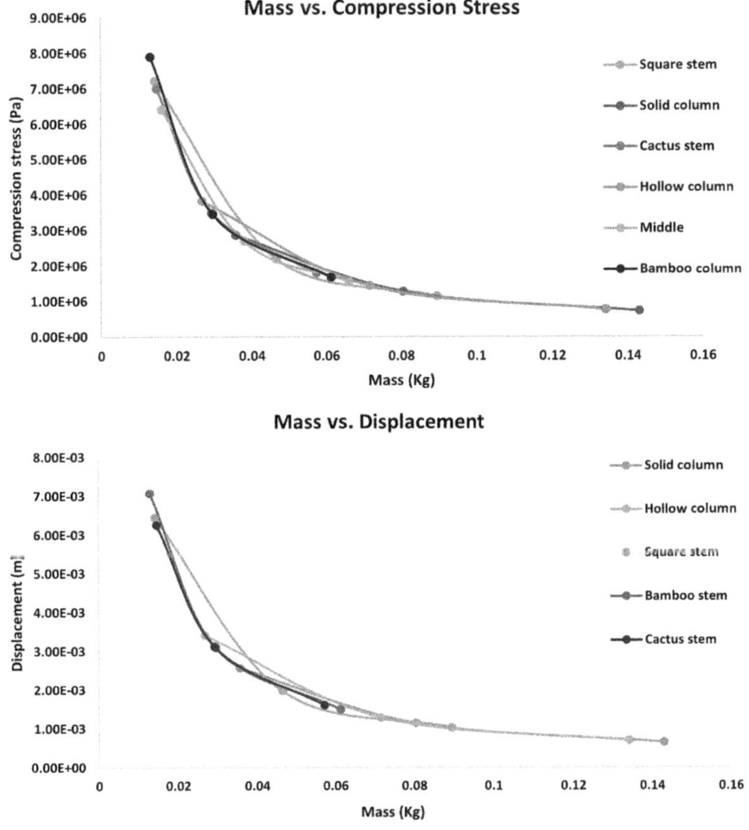

FIGURE 4.9 (a) compressive stress and (b) axial displacement of optimized rods with respect to mass.

4.8 CONCLUSIONS

Biomimicry, the practice of imitating natural processes, components, and architectures, offers a rich source of inspiration for engineering innovations. By studying natural elements and their evolutionary adaptations, researchers aim to create artificial counterparts with enhanced load-carrying capacity and functionality. This chapter focuses on biomimetic approaches to designing structural elements, particularly slender columns, to improve their resistance to buckling, a common failure mode in load-bearing structures.

Researchers draw inspiration from natural structures such as plant stalks and roots, which exhibit porous architectures that enhance their buckling resistance. To mimic these natural designs, digital representations of biological structures are created and analyzed using machine learning algorithms. These algorithms identify correlations between biological features and buckling load, enabling the optimization of biomimetic designs.

Data generation and fingerprinting techniques are employed to create a diverse set of biomimetic rods with varying shapes and internal structures. MATLAB code is used to generate and manipulate structural features, resulting in thousands of biomimetic rod designs. Each design is assigned a unique fingerprint, allowing for efficient organization and analysis of the dataset.

Supervised machine learning algorithms are then trained on the dataset, using input variables such as mass, volume, and microstructure features to predict buckling load. The Ensemble Bagged Tree algorithm emerges as the most effective for this task, outperforming other models such as Support Vector Machines (SVM) and Gaussian Process Regression (GPR).

The Discovery of new biomimetic rods with superior buckling loads is identified through further filtering of the dataset. These designs are converted into 3D CAD models and analyzed using finite element analysis to validate their structural properties. A total of 160 new biomimetic rods with buckling loads higher than that in the training dataset were discovered. The results demonstrate that the newly discovered biomimetic rods exhibit significantly higher buckling strength compared to the original biomimetic designs.

Overall, this study highlights the effectiveness of biomimetic approaches combined with machine learning techniques in optimizing structural elements for improved performance and resilience against buckling.

REFERENCES

1. Vincent JFV, Olga AB. Materials in nature. Mater. Today, 6:37–42, (2003).
2. Lee H, Lee BP, Messersmith PB. A reversible wet/dry adhesive inspired by mussels and geckos. Nature, 448(7151):338, (2007).
3. Autumn K, Sitti M, Liang YCA, Peattie AM, Hansen WR, Sponberg S, Kenny TW, Fearing R, Israelachvili JN, Full RJ. Evidence for van der Waals adhesion in gecko setae. Proc. Nat. Acad. Sci. United States of America, 99:12252–12256, (2002).
4. Liu YW, Wang H, Li JC, Li PY, Li SJ. Gecko-inspired controllable adhesive: Structure, fabrication, and application. Biomimetics, 9:149, (2024).

5. Olender J, Young C. Examination of gecko-inspired dry adhesives for heritage conservation as an example of iterative design and testing process for new adhesives. European Phys. J. Plus, 138:644, (2023).

6. Ha NS, Lu GX, Xiang XM. Energy absorption of a bio-inspired honeycomb sandwich panel. J. Mater. Sci., 54:6286–6300, (2019).

7. Gibson LJ. Woodpecker pecking: How woodpeckers avoid brain injury. J. Zool., 270:462–465, (2006).

8. Sandeep CS, Evans TM. Biomimetic intruder tip design for horizontal penetration into a granular pile. Bioinspiration & Biomimetics, 18:064001, (2023).

9. Zhang WP, Li RA, Yang QZ, Fu Y, Kong XQ. Impact resistance of a fiber metal laminate skin bio-inspired composite Sandwich panel with a rubber and foam dual core. Materials, 16:453, (2023).

10. Liu YZ, Qiu XM, Yu TX, Tao JW, Cheng Z. How does a woodpecker work? An impact dynamics approach. Acta Mech. Sinica, 31:181–190, (2015).

11. Tang ZY, Kotov NA, Magonov S, Ozturk B. Nanostructured artificial nacre. Nat. Mater., 2:413, (2003).

12. Mao LB, Gao HL, Yao HB, Liu L, Cölfen H, Liu G, Chen SM, Li SK, Yan YX, Liu YY, Yu SH. Synthetic nacre by predesigned matrix-directed mineralization. Science, 354:107–110, (2016).

13. Sellinger A, Weiss PM, Nguyen A, Lu YF, Assink RA, Gong WL, Brinker CJ. Continuous self-assembly of organic-inorganic nanocomposite coatings that mimic nacre. Nature, 394:256–260, (1998).

14. Manoonpong P, Patanè L, Xiong XF, Brodoline I, Dupeyroux J, Viollet S, Arena P, Serres JR. Insect-inspired robots: Bridging biological and artificial systems. Sensors, 21:7609, (2021).

15. Dixit S, Stefanska A. Bio-logic, a review on the biomimetic application in architectural and structural design. AIN SHAMS Eng. J., 14:101822, (2023).

16. Wegst UGK, Bai H, Saiz E, Tomsia AP, Ritchie RO. Bioinspired structural materials. Nat. Mater., 14:23–36, (2015).

17. Munch E, Launey ME, Alsem DH, Saiz E, Tomsia AP, Ritchie RO. Tough, bio-inspired hybrid materials. Science, 322:1516–1520, (2008).

18. Gao HJ, Ji BH, Jäger IL, Arzt E, Fratzl P. Materials become insensitive to flaws at nanoscale: Lessons from nature. Proc. Nat. Acad. Sci. United States of America, 100:5597–5600, (2003).

19. Bhushan B. Biomimetics: Lessons from nature-an overview. Philos. Trans. R. Soc. A Math. Phys. Eng. Sci., 365:9–24, (2007).

20. Ahamed MK, Wang HX, Hazell PJ. From biology to biomimicry: Using nature to build better structures-a review. Constr. Build. Mater., 320:126195, (2022).

21. Da DC, Qian XP. Fracture resistance design through biomimicry and topology optimization. Extreme Mech. Lett., 40:100890, (2020).

22. Sanga R, Kilumile M, Mohamed F. Alternative clay bricks inspired from termite mound biomimicry. Case Stud. Constr. Mater., 16:e00977, (2022).

23. Nor MSAM, Aliff M, Samsiah N. A review of a biomimicry swimming robot using smart actuator. Int. J. Adv. Comput. Sci. Appl., 12:395–405, (2021).

24. Kumar A, AL-Jumaili A, Bazaka O, Ivanova EP, Levchenko I, Bazaka K, Jacob MV. Functional nanomaterials, synergisms, and biomimicry for environmentally benign marine antifouling technology. Mater. Horiz., 8:3201–3238, (2021).

25. Barba A, Diez-Escudero A, Espanol M, Bonany M, Sadowska JM, Guillem-Marti J, Öhman-Mägi C, Persson C, Manzanares MC, Franch J, Ginebra MP. Impact of biomimicry in the design of osteoinductive bone substitutes: Nanoscale matters. ACS Appl. Mater. Interf., 11:8818–8830, (2019).

26. du Plessis A, Babafemi AJ, Paul SC, Panda B, Tran JP, Broeckhoven C. Biomimicry for 3D concrete printing: A review and perspective. Addit. Manuf., 38:101823, (2021).

27. Eadie L, Ghosh TK. Biomimicry in textiles: Past, present and potential. An overview. J. R. Soc. Interf., 8:761–775, (2011).

28. Nouri A, Zarkesh A. Efficient configuration in architectural structures based on biomimicry principles in femur bone using hurricane geometry. Innov. Infrastruct. Solut., 9:53, (2024).

29. Bhushan B. Biomimetics: Lessons from nature – an overview. Philos. Trans. Roy. Soc. A Math. Phys. Eng. Sci., 367:1445–1486, (2009).

30. Nji J, Li G. A biomimic shape memory polymer based self-healing particulate composite. Polymer, 51:6021–6029, (2010).

31. Li G, Nettles D. Thermomechanical characterization of a shape memory polymer based self-repairing syntactic foam. Polymer, 51:755–762, (2010).

32. Li G, Uppu N. Shape memory polymer based self-healing syntactic foam: 3-d confined thermomechanical characterization. Compos. Sci. Techno., 70:1419–1427, (2010).

33. Ross JL. Autonomous materials from biomimicry. MRS Bullet., 44:119–123, (2019).

34. Mayer G. Rigid biological systems as models for synthetic composites. Science, 310:1144–1147, (2005).

35. Yang Y, Song X, Li XJ, Chen ZY, Zhou C, Zhou QF, Chen Y. Recent progress in biomimetic additive manufacturing technology: From materials to functional structures. Adv. Mater., 30:1706539, (2018).

36. Shyy W, Kang CK, Chirarattananon P, Ravi S, Liu H. Aerodynamics, sensing and control of insect-scale flapping-wing flight. Proc. R. Soc. A Math. Phys. Eng. Sci., 472:20150712, (2016).

37. Li JL, Huang ZR, Liu G, An QL, Chen M. Topology optimization design and research of lightweight biomimetic three-dimensional lattice structures based on laser powder bed fusion. J. Manuf. Process, 74:220–232, (2022).

38. Vogel S. Comparative Biometeorology: Integrated Approaches in Adaptation and Acclimation. Springer Science & Business Media, (2013).

39. Meyers MA, Mckittrick J, Chen PU. Structural biological materials: Critical mechanics-materials connections. Science, 339:773–779, (2013).

40. Zhao N, Wang Z, Cai C, Shen H, Liang F, Wang D, Wang C, Zhu T, Guo J, Wang Y, Liu X, Duan C, Wang H, Mao Y, Jia X, Dong H, Zhang X, Xu J. Bioinspired materials: From low to high dimensions. Adv. Mater., 33:2100234, (2021).

41. Espinosa HD, Seunghwa R. A review on mechanical properties of bamboo and bio-based composites. Front. Mater., 7:51, (2020).

42. Pohl G. Biomimetics for Architecture: Learning from Nature. Springer Science & Business Media, (2011).

43. Theraulaz G, Bonabeau E. A brief history of stigmergy. Artif. Life, 5:97–116, (1999).

44. Seidel R, Gerhard MA. On the structural design of trees. J. Theor. Biol., 210:171–183, (2001).

45. Gibson LJ, Ashby MF. Cellular Solids: Structure and Properties. Cambridge University Press, (1997).

46. Barthelat F, Botsis P. Biological materials and structures for impact protection. J. R. Soc. Interf., 7:19–34, (2010).

47. LeCun Y, Bengio Y, Hinton G. Deep learning. Nature, 521:436–444, (2015).

48. Schmelzle S. Deep learning for materials science. Adv. Intell. Syst., 2:1900122, (2020).

49. Yan C, Feng X, Li G. From drug molecules to thermoset shape memory polymer: A machine learning approach. ACS Appl. Mater. Interf., 13:60508–60521, (2021).

50. AlAli M, Mattar Y, Alzaim MA, Beheiry S. Applications of biomimicry in architecture, construction and civil engineering. Biomimetics, 8:202, (2023).

51. Dottore ED, Mondini A, Rowe N, Mazzolai B. A growing soft robot with climbing plant–inspired adaptive behaviors for navigation in unstructured environments. Sci. Robot., 9:86, (2024).

52. Yoseph BC. Biomimetics: Biologically Inspired Technologies. CRC Press, (2006).

53. Benyus JM. Biomimicry: Innovation Inspired by Nature. Harper Collins, (2009).

54. Ultimaker BV, Ultimaker Cura, version 4.7.1,

55. Utrecht S, Kovan V, Altan G, Topal ES. Effect of layer thickness and print orientation on strength of 3D printed and adhesively bonded single lap joints. J. Mech. Sci. Technol., 31:2197–2201, (2017).

56. Aru M, Ghanshyam P, Tran DH, Turab L, Rami R. Machine learning strategy for accelerated design of polymer dielectrics. Sci. Rep., 10:20952, (2016).

57. Wu S, Kondo Y. Machine-learning-assisted discovery of polymers with high thermal conductivity using a molecular design algorithm. Comput. Mater., 5:66, (2019).

58. Cao YF, Wu W, Zhang HL, Pan JM. Prediction of the elastic modulus of self-compacting concrete based on SVM. Trans Tech Publications, 357:1023–6, (2013).

59. Chen H, Qian C, Liang C, Kang W. An approach for predicting the compressive strength of cement-based materials exposed to sulfate attack. PLoS One, 13:0191370, (2018).

60. Salehia H, Burgueño R. Emerging artificial intelligence methods in structural engineering. Eng. Struct., 171:170–189, (2018).

61. Liang L, Liu M, Martin C, Sun W. A deep learning approach to estimate stress distribution: A fast and accurate surrogate of finite-element analysis. J. R. Soc. Interf., 15:20170844, (2018).

62. Qi Z, Zhang N, Liu Y, Chen W. Prediction of mechanical properties of carbon fiber based on cross-scale FEM and machine learning. Compos. Struct., 212:199–206, (2019).

63. Capuano G, Julian J. Smart finite elements: A novel machine learning application. Comput. Methods Appl. Mech. Eng., 345:363–384, (2019).

64. Matrinez F, Moreno MJ, Matrinez M, Solves JA, Loreta D. A finite element-based machine learning approach for modeling the mechanical behavior of the breast tissues under compression in real-time. Comput. Biol. Med., 90:116–124, (2017).

65. Smola A, Vishwanathan SVN. Introduction to Machine Learning. Cambridge University Press, (2010).

5 Structural Optimization of Lattice Structures

5.1 LATTICE STRUCTURES

A lattice structure is formed by stacking lattice unit cells cheek by jowl in any desired order. The entire lattice's structural performance depends on the construction of the lattice unit cell it is made of. Extensive research has been carried out in the design, fabrication, and evaluation of these lightweight architectures. Depending on the number of struts and number of joints in a single unit cell, it can be classified as a stretching- or bending-dominated structure [1]. In a stretching-dominated structure, the primary mode of failure is through stretching of the struts, while in a bending-dominated structure, the primary mode of failure is through bending of the struts. Compared to foam and bending-dominated lattice structures, stretching-dominated lattice structures were proven to perform better in terms of strength and stiffness [1]. Several lattice unit cells were proposed with superior performance and various advantages in structural, thermal, impact, vibrational, and acoustic domains [2–6]. Octet lattice structure is one of the best truss-based lattice structures among stretching-dominated lattice unit cells [7]. Gyroid and double gyroid structures manufactured through additive manufacturing exhibited decent impact absorption capabilities and additional advantages with the same stiffness in all the axial directions [8]. Hollow micro-truss lattice structures were studied for their enhanced energy absorption [9]. Hybrid sandwich panels made of pyramid truss structures as cores were fabricated with higher damping performance [10]. Several numerical and experimental studies were conducted by various groups to reinforce the proposed structures. Linear and nonlinear effective properties of lattice structure were studied using continuum theory models. Deshpande and Fleck initially studied the effective properties of octet lattice structure [7].

The fabrication techniques and structural performance of lattice-cored sandwich structures were explored by different groups [11–14]. Selective laser melting (SLM) is used to manufacture bioinspired Kagome sandwich panels with titanium core. The compression and shear properties of these sandwich structures are evaluated and proved to perform better than honeycomb aerospace core structures [11]. Stereolithography (SLA) is used to print several lattice cores made of epoxy-based photopolymer resin and sandwich plates made of carbon-fiber-reinforced face sheets which are co-cured in a two-stage manufacturing process. The compression and flexural properties of the sandwich structures were evaluated [12]. The bending response of graded lattice core sandwich structures is evaluated which achieves minimum weight sandwich structures [13]. Silicon rubber molds

DOI: 10.1201/9781003400165-5

were used to fabricate CFRP tetrahedral core sandwich structures and tested for compression and shear strengths [14]. It is known that the lattice core in sandwich structures plays an important role in the overall load-carrying capacity of the sandwich. With the advancement in 3D printing, lattice cores with overly complex geometrical configurations can be manufactured. Therefore, further improvement in the load-carrying capacity in lattice-cored sandwich depends on the optimization of lattice cores.

Topology optimization technique was used to further optimize existing lattice unit cells and propose unique structures [15–24]. In this technique, a previously available unit cell was improved iteratively for better performance by keeping the relative density constant each time [15]. Through this technique, new optimal lattice unit cells (Octahedral and Rectified Cubic (ORC) and an oblate and quasi-spherical octahedral (OQSO)) which were 5% and 38% stiffer compared to octet lattice unit cells in standard (001) direction were developed [15]. To compensate for the elastic anisotropic nature of octet lattice unit cells, researchers developed elastically isotropic unit cells by merging different basic unit cells such as simple cubic unit cells, octet unit cells, etc. [16]. Optimization techniques were developed to design and automatically generate truss structures within given boundary conditions [17]. Messner [18] proposed an inverse optimization approach and optimized an iso-truss lattice, which showed about a 50% increase in stiffness over the octet lattice. In recent years, researchers have used topology optimization to design and optimize lattice structures for improved mechanical properties, such as strength and energy absorption. For example, Ang and Li [17] designed a lattice structure inspired by cuttlebone using topology optimization and found that it had a 141.96% increase in relative collapse strength compared to the standard octet lattice. Other researchers, such as Yang et al. [19], have also used topology optimization to optimize the density distribution and topology of lattice structures for improved mechanical performance. However, classical topology optimization can be challenging to implement, especially for lattice structures with specific constraints or boundary conditions. This is due to the complex two-stage genetic algorithm (GA) coding involved and the reliance on mass reduction, which can overlook structures that show significant strength improvements with small mass additions. The multiple iterations required to optimize a structure and the need to analyze each new structure using auxiliary software make the process time-consuming and computationally intensive. Furthermore, topology optimization can only produce a few improved structures compared to a given reference design. It is noted that there is a rich supply in the literature on the optimization of lattice structures. Many new lattice structures and manufacturing methods have been proposed over the years [25–38]. Readers who are interested in learning more about the recent development in lattice structures are encouraged to review this comprehensive literature.

Although considerable progress has been made in proposing new optimal lattice unit cells, it is believed that there is still a vast volume of unexplored design space in lattice unit cell designs that could perform significantly better compared to the structures that have been proposed so far. This situation presents an opportunity for machine learning to assist in exploring the uncharted territory, bypassing the need for complex numerical analysis. Furthermore, machine learning can help overcome the limitations of classical topology optimization or other optimization techniques that require significant time and computational power for large datasets.

Recent advances in machine learning have enabled the design of auxetic metamaterials [39–43], spinodal metamaterials [44–46], and 3D chiral metamaterials [47–50] with improved properties while reducing the analytical and computational burden. For instance, neural networks and inverse design techniques have been used to design spinodal metamaterials with higher anisotropic stiffness, while auxetic metamaterials have been designed using machine learning to circumvent complex analytical and computational load. Similarly, machine learning has been applied to design and optimize 3D chiral metamaterials with strong chiroptical response, thus avoiding the need for time-consuming numerical simulations. However, to date, no study has focused on discovering new lightweight lattice truss structures and exploring the design space of such structures using machine learning.

To explore and identify optimal lattice structures for uniaxial compression analysis in different loading orientations, we generated a massive dataset of fingerprints of lattice unit cells. To accomplish this, a representative volume element (RVE) was designed in such a way that it could generate various combinations of unit lattice structures by connecting different points. The dataset included both stretching- and bending-dominated lattice structures. However, conventional numerical and simulation methods would require enormous human power, time, and computational processing power to evaluate, segregate, and identify optimal structures from this vast ocean of data.

The objective of this chapter is to design optimal stretch- and bending-dominated lattice unit cells through machine learning. To this end, this study was conducted by training an adequate portion of the generated dataset consisting of fingerprints of lattice unit cells for uniaxial compression analysis in different loading orientations. A RVE made of several points was considered in this study. It was designed in such a way that a simple MATLAB code could be used to generate various possible combinations of unit lattice structures by connecting different points in the RVE. This included both stretching- and bending-dominated lattice structures. With this, a huge dataset with combinations of various mechanisms and structures was generated. To evaluate, segregate, and identify optimal structures from this ocean of data, conventional numerical and simulation methods will require enormous human power and time and computational processing power and still will not be optimal. To handle this issue, machine learning which is a very powerful tool in terms of computational speed and data handling can be employed. Regression training with an adequate amount of input data with machine learning models like support vector regression (SVR), decision trees, Gaussian process regression (GPR), random forest, etc., can be used to predict the stresses, deformation, and strengths in a lattice structure. Similar classification training can be used to extract several desired features from a given dataset [51]. ANSYS designer module and simulation tool were used to generate the fingerprints.

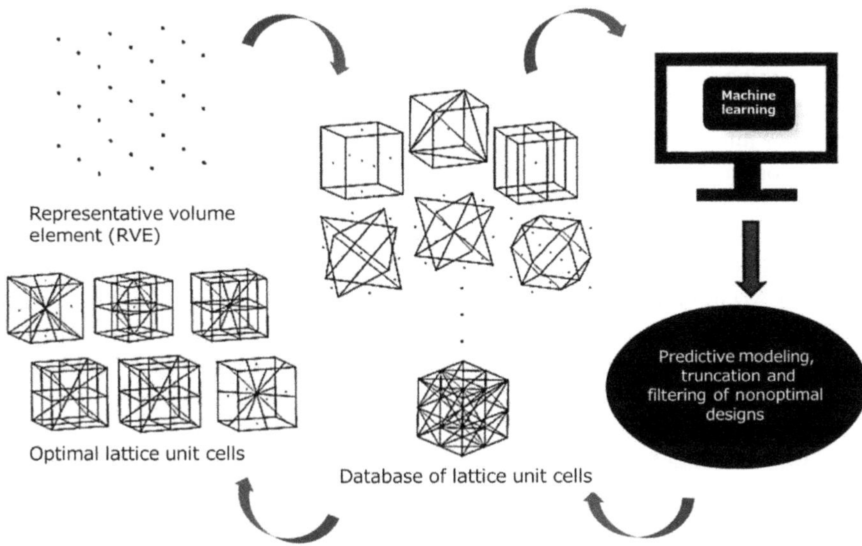

FIGURE 5.1 Schematic flowchart of the machine learning process to discover new lattice unit cells.

This generated dataset was used to train machine learning algorithms to evaluate and predict the mechanical properties of other untrained lattice unit cells. By comparing with octet lattice unit cells, optimal lattice structures were identified. Figure 5.1 shows the schematic of the machine learning process. Simulation and experimental validations were conducted for uniaxial compression with respect to relative density on several optimal lattice unit cells at different orientations. Modeling conditions like the overall volume, diameter of individual truss elements, material properties, mesh sizing, boundary conditions, etc. were maintained constant for all designs for systematic comparisons. To further enhance the optimized unit cells, the solid elements in the unit cells were replaced by porous biomimetic rods, leading to significantly higher buckling resistance. In the following, we will discuss in more detail each step involved in discovering new lattice unit cells using machine learning.

5.2 DATA GENERATION AND FINGERPRINTING OF LATTICE UNIT CELLS

To generate all possible lattice unit cell combinations, a consistent approach is necessary, which requires certain boundary conditions to be set. In this study, a cuboid consisting of 27 points was used, which can be divided into eight symmetric small cubes, with eight points at the edges of each cube. By connecting each point to its neighboring point, single lattice elements are formed. By combining multiple lattice elements, a wide range of three-dimensional isotropic and anisotropic lattice unit cells can be generated. To simplify data size and reduce complexity in handling data, one-eighth of the entire cube is considered as a RVE to generate all possible symmetric lattice unit cells. This RVE can be rotated in all remaining seven mirror images to build the entire lattice unit cell.

Furthermore, by considering the entire cube (27 points) as the RVE, several other anisotropic combinations can be generated to produce structures that may be optimal in direction-dependent loading conditions. When designing the training dataset using the ANSYS workbench, a constant truss diameter, material parameters, meshing size, and boundary conditions are set for all the fingerprints or lattice structures to maintain consistency. A MATLAB function called "combnk" is used to enumerate all possible combinations using the RVE. A total of 176 individual elements can be formed within the RVE, and unique designs and lattice unit cells covering structures consisting of from four to all 176 elements can be generated. In this way, nearly one million unique lattice unit cell designs can be formed, which covers a vast design space.

Now, we would like to provide more details on the RVE, the creation of the unit cells, and the fingerprinting of the unit cells. The first step is to create a RVE; see Figure 5.2. The one-eighth part of the RVE can be used to connect different combinations of points to form all possible combinations of symmetric and asymmetric lattice unit cells. Here, each point of the entire RVE is numbered from 1 to 27 (Figure 5.2(a)). One-eighth part of the RVE consists of eight points numbered from 1 to 8 (Figure 5.2(b)). Each pair of numbers represents the

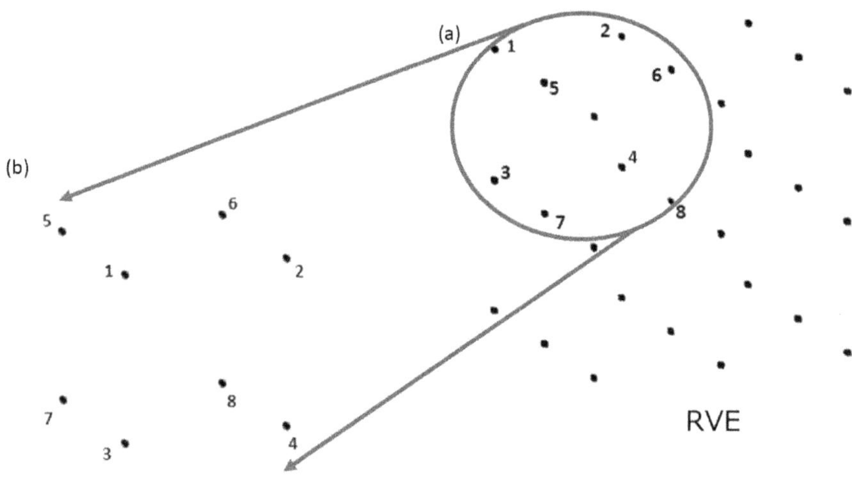

FIGURE 5.2 (a) Representative volume element (RVE) and (b) one-eighth part of the RVE which can be used to form various combinations of symmetric lattice structures. Procedure to form different lattice unit cells based on RVE. The figure consists of a 3 by 3 matrix of points in 3 dimensions. It also magnifies a 2 by 2 section of the 3 by 3 matrix to name each of the points from 1 to 8 representing one-eighth of an RVE.

element connecting the respective points. The line connecting the corresponding points is abstracted as the rod in the lattice unit cell. To maintain consistency and meaningful connectivity, a condition is set which restricts the formation of elements to connect only adjacent points, i.e., point 1 can only form an element by connecting to points 2, 3, 4, 5, 6, 7, and/or 8, which are immediately adjacent to point 1, but not with any other points from points 9 to 27 because there will be another point in between point 1 and points 9–27. In such a way, each element corresponds to a unique pair of numbers, which is the fingerprint of the lattice element. For example, (12) is a fingerprint representing an element formed by connecting points 1 and 2. Likewise, all the elements in the RVE can be represented by the two points they connect. To form symmetric lattice structures, the primary and complementary elements shown in Table 5.1 are enough. Here, the primary elements refer to the elements used to name each symmetric lattice unit cell fingerprint. For example, (12 24 46) is a symmetric lattice unit cell. Points 12, 24, and 46 are the elements from the one-eighth part of the 27-point cuboid used to represent the (12 24 46) lattice unit cell as a fingerprint. To maintain symmetry, the complementary elements ((13, 15) for (12), (26, 34, 37, 56, 57) for (24), and (47, 67) for (46)) attached to each primary element in Table 5.1 are essential. Now, by using the "combnk" function, various combinations of primary elements can be generated. Combining with the complementary elements and mirror rotations, symmetric lattice unit cells can be designed.

To form direction-dependent asymmetric lattice unit cells, the cuboid with 27 points is considered as RVE. Here, Table 5.1 cannot be used as asymmetric lattice unit cells which are formed by combining elements randomly within the RVE. Hence, the number of elements to form a particular lattice unit cell is predefined, and various combinations are generated using the same "combnk" function. For example, (12 13 15 213 1314 1415 1015 910 39 519 1316 1625 1518 1827 911 1121 1922 2225 2526 2627 1920 2021 2124 2427 18 89 813 815 819 821 825 82725

TABLE 5.1

Primary and Their Complementary Elements to Form Symmetric Lattice Unit Cells

Primary Elements	Complementary Elements				
12	13	15			
16	14	17			
24	34	37	56	57	26
25	23	35			
28	38	58			
46	47	67			
48	68	78			
45	27	36			
18					

23 35 28 58 46 47 67 46 68 78 45 27) is an asymmetric lattice unit cell formed by the 43 elements represented in the fingerprint. Here, elements with two digits (12, 13, 15, etc.) mean the element is formed by connecting two points; the first digit represents one point, and the second digit represents the other point. For the elements with three digits (213, 519, 911, etc.), the first digit represents one point, and the remaining two digits are the number of the other point. In the case of elements with four digits (1314, 1415, 1015, etc.), the first two digits represent one point, and the remaining two digits represent the other point forming the element. The reason that the fingerprints for some rods need three digits and for some rods need four digits is that for the RVE with 27 points, points 1–9 are represented by one digit, while points 10–27 are represented by two digits. Therefore, if point 2 is connected to point 12, the fingerprint is (212), and if point 10 is connected to point 25, the fingerprint is (1025). For lattice structures or lattice-cored sandwich structures, they are formed by placing a lattice unit cell side by side in rows and columns. To further visualize the fingerprint, Figure 5.3 shows a cubic lattice unit cell. The fingerprinting for the lattice unit cell in Figure 5.3 can be represented by (12 13 15 24 26 213 37 39 34 410 414 48 56 57 519 68 616 622 711 78 812 817 823 910 911 1012 1015 1112 1121 1218 1224 1314 1316 1415 1417 1518 1625 1617 1718 1726 1827 1922 1920 2021 2023 2124 2223 2225 2324 2326 2427 2526 2627). Here, 12 refers to the lattice element connecting vertices 1 and 2; similarly, 13 refers to the

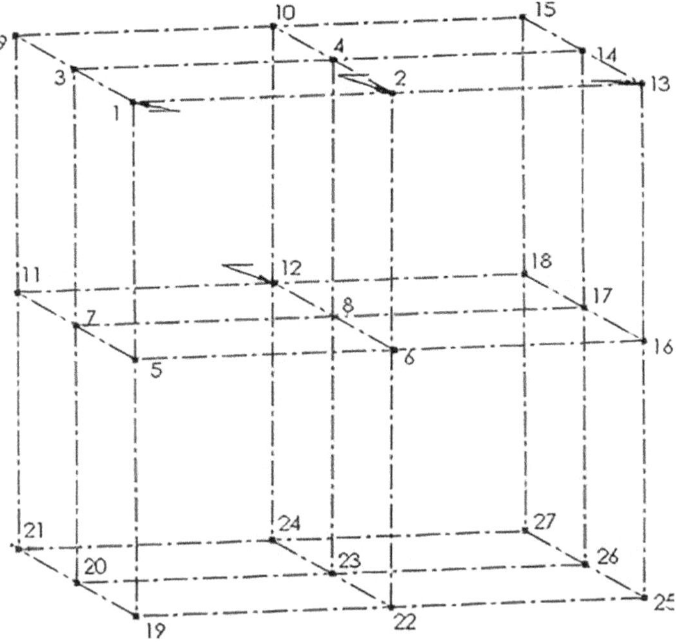

FIGURE 5.3   Sample cubic lattice unit cell. The vertices are named from 1 to 27, and the dotted lines connecting two adjacent vertices are the lattice elements or rods.

lattice elements 1 and 3 and so on. The naming of the element does not depend on the sequence, i.e., the element connecting vertices 1 and 2 can be named as 12 or 21. Since 12 does not hold any value; it will be considered as a categorical variable by the machine learning algorithms.

To generate the training dataset, we also need the mechanical properties of the created lattice unit cells, which were obtained by conducting finite element analysis. In this study, the ANSYS designer module and simulation tool were used, and the generated dataset was used to train machine learning algorithms to evaluate and predict the mechanical properties of other untrained lattice unit cells. By comparing with the octet lattice unit cell, optimal lattice structures were identified through simulation and experimental validations for uniaxial compression with respect to relative density on several optimal lattice unit cells at different orientations.

Modeling conditions like the overall volume, diameter of individual lattice elements, material properties, mesh sizing, boundary conditions, etc. were kept constants for all designs to facilitate systematic comparisons. To further enhance the optimized unit cells, the solid elements in the unit cells were replaced by porous biomimetic rods discovered in Chapter 4, leading to significantly higher buckling resistance.

5.3 MACHINE LEARNING

Once the created lattice unit cells are fingerprinted and their mechanical properties are obtained using finite element analysis, we can use machine learning to establish the correlation between fingerprints and mechanical properties such as compressive strength. As mentioned earlier, machine learning can be used both for classification and regression. Machine learning uses the regressing tendencies from the training dataset to understand the hidden relation between different data points. Statistical models will be used to establish this relationship, which can perform the desired regression or classification exercise. Machine learning models like kernel ridge regression (KRR), GPR, neural networks (NNs), and SVM have been used as forward models to predict polymer properties based on chemical characteristics of various polymers and discover new polymer combinations with enhanced properties [51–57]. KRR model was used to predict the electronic dielectric, ionic dielectric, and bandgap properties of polymers which was used to bypass the density functional theory (DFT) calculations with an average prediction error of 10% or less. Based on the polymer building block identities, fingerprints (numerical representation) of each polymer were built, which were used for the on-demand property prediction using the KRR model [51]. Later GA approach was used to search materials whose properties were easily predicted using the ML model, avoiding complex numerical and experimental testing [52]. Neural networks and random forest models were trained to estimate the glass transition temperature, melting temperature, and thermal conductivity of polymers. The thermal conductivity models exhibited a mean absolute error (MAE) of 0.0204. Similar to the previous models, once a forward

ML model was developed, a molecular design algorithm was used to generate several new samples. By using the ML models to predict the desired properties, new polymers with thermal conductivities of 0.18–0.41 W/mK were discovered [52]. A convolutional neural network (CNN) was used to discover new shape memory polymers with higher recovery stress [56, 57]. SVM models were used to predict the elastic modulus of self-compacting concrete with an MAE of less than 5% as an alternative to other numerical models whose results were often ambiguous and uncertain [58].

It can be observed from the literature that with data having logical interrelation between the predictors and the responses, machine learning tool can be more accurate and requires less data. Hence, machine learning can be adequately used for the prediction of mechanical properties in structural engineering. In this study, a training dataset consisting of about 2,000 random lattice unit cells is extracted for regression training from the entire dataset generated using MATLAB. ANSYS modeling tool is used to analyze each combination. Uniaxial compression is performed with loads normal to the Cartesian coordinate axis and at angular orientations. All the models are subjected to a constant load, the same overall volume, and material properties. While other high-performance thermoset polymers are available for 3D or 4D printing [59, 60], the commercially available RSI-10 photopolymer is used as the base material throughout this research. The mass, elemental compression stress, tensile stresses, and deformation are recorded for each design to form a dataset for the regression training. Coherent data management is crucial for precise predictions in any machine learning process. In this study, the input data are the fingerprints of each design. Fingerprints refer to the numerical representation of the individual lattice unit cells, as discussed in the previous section. There is no definitive procedure for fingerprinting the lattice unit cells for machine learning. Depending on the type of input data, fingerprints should be defined as a logical numerical sequence maintaining consistency throughout the dataset. To fingerprint each lattice unit cell in this study, each lattice point in the RVE is numbered, and the element connecting any two points is fingerprinted by the numbers of the two points that the element connects. For symmetric lattice unit cells, the one-eighth part of the 27-point cuboid is considered as RVE. For example, (12 24 46) lattice unit cell represents three elements connecting points 1 and 2 (element 12), 2 and 4 (element 24), and 4 and 6 (element 46) in the one-eighth section. Since it is a symmetric unit cell, these three elements shall be rotated into all the remaining seven mirror planes of the 27-point cuboid to form the entire lattice unit cell. In the case of asymmetric lattice unit cells, the 27-point cuboid is considered as RVE, and all the elements present in the lattice unit cell together form the fingerprint (Figure 5.2). Keeping the fingerprints as the predictors (input) and desired features like mass, uniaxial compression along principal orientation, and uniaxial compression with angular load orientation as response variables (outputs), different datasets are created individually. These datasets can be used to estimate the structural and mechanical properties of the remainder of the fingerprints without any further simulation or analysis. MATLAB is used to train the machine learning algorithms and predict the individual lattice unit cell characteristics. Each dataset is trained and tested with multiple machine learning algorithms to find the most suitable method.

In this study, the machine learning algorithms were trained and tested using MATLAB, and rational quadratic GPR produced optimal results compared to SVMs, ensemble methods, and decision trees. GPR is a Bayesian approach to regression, which means that it is based on Bayes' theorem. Bayes' theorem is a way to update the probability of a hypothesis based on new evidence. In the context of GPR, the hypothesis is the function that relates the input variables (the fingerprints in this case) to the output variables (the mechanical properties).

In Bayesian GPR, we start with a prior distribution over the possible functions that could relate the inputs to the outputs. This prior distribution captures our beliefs about what kinds of functions are likely to be good models for the data. Then, we use the observed data to update this prior distribution to obtain a posterior distribution over the functions. The posterior distribution is typically a Gaussian process, which means that it is a distribution over functions rather than a distribution over parameters. The Gaussian process is characterized by a mean function and a covariance function. The mean function represents the most likely function given the data, and the covariance function represents the uncertainty in our predictions. The parameters of the Gaussian process are typically learned from the data using maximum likelihood estimation. This involves finding the values of the parameters that make the observed data most likely under the Gaussian process model, making it a suitable choice for this study. While performing machine learning, rational quadratic GPR produced optimal results compared to SVMs, ensemble methods, and decision trees. This conclusion is made based on the root-mean-square error (RSME) values of these models. A list of comparisons of several machine learning regression techniques and their performances can be found in Figure 5.4 and Table 5.2. GPR is a Bayesian approach to regression that works well for small datasets [61]. Therefore, the use of machine learning algorithms to predict mechanical properties in structural engineering, especially for small datasets, can be highly effective, and this study provides a framework for this approach.

Figure 5.4 exhibits the response plots contrasting the anticipated and factual outcomes of diverse properties through the usage of GPR. The outcomes obtained via machine learning are indeed in considerable accordance with the true values stemming from ANSYS simulation outcomes. The forecast line and observation dots delineate the relationship between the predictions and observed values, whereby the proximity of the latter to the former denotes the efficiency of the model. Upon completion of forecasting the mechanical characteristics of the dataset excluded from the training process utilizing the GPR models, said prognostications are subsequently juxtaposed with the performance of the octet lattice unit cell scrutinized under analogous boundary conditions. In accordance with this methodology, a cumulative of 20 optimal unit cells for symmetric lattice unit cells that outmatch the mechanical properties of the octet lattice unit cell is suggested.

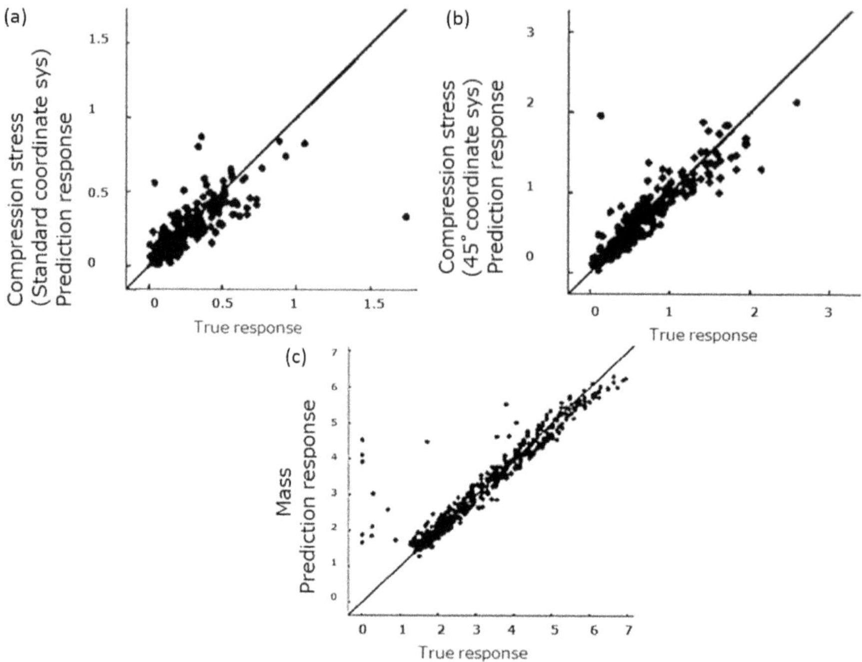

FIGURE 5.4   True response by ANSYS finite element analysis versus prediction by GPR machine learning algorithm for (a) uniaxial compression stresses in Cartesian coordinate system, (b) 45° orientation compression stress, and (c) mass.

Table 5.2 provides a comparison of diverse machine learning regression models used to predict compression stress and mass. MATLAB, a widely used software program, is utilized to carry out the regression analysis. MATLAB has a built-in regression learner application that facilitates the direct import of both training and testing datasets to different machine learning algorithms. To ensure the accuracy and validity of the results, a fivefold cross-validation method was implemented across all regression models. The RMSE is a commonly used metric to evaluate the performance of a machine learning regression model. It measures the average magnitude of the error between the predicted and actual values of the target variable. RMSE is calculated by taking the square root of the mean of the squared differences between the predicted and actual values.

In simpler terms, RMSE is a way to determine how far off the predictions of a regression model are from the actual values. It considers both the magnitude and direction of the errors and provides a single value to summarize the overall performance of the model. A lower RMSE value indicates a better fit of the model to the data. To calculate the RMSE, the differences between each predicted value and its corresponding actual value are first squared, then summed up, and divided by the

TABLE 5.2

Comparison of Several ML Models

Uniaxial Compression (Standard Coordinate System)

Machine Learning Technique	Training Dataset			Testing Dataset		
	RMSE	R^2	MAE	RMSE	R^2	MAE
Rational quadratic GPR	0.20469	0.92	0.10348	0.522207	NA	0.436
Ensemble	0.24138	0.88	0.11337	0.278609	NA	0.251
Cubic SVM	0.23717	0.89	0.13625	0.433322	NA	0.339
Fine tree	0.33009	0.78	0.13227	0.522207	NA	0.436

Uniaxial Compression (45° Coordinate System)

Machine Learning Technique	Training Dataset			Testing Dataset		
	RMSE	R^2	MAE	RMSE	R^2	MAE
Rational quadratic GPR	0.26186	0.94	0.12043	0.2847	NA	0.221
Ensemble	0.28321	0.93	0.13955	0.3347	NA	0.251
Cubic SVM	0.28385	0.93	0.15415	0.45781	NA	0.348
Fine tree	0.32882	0.91	0.13205	0.51247	NA	0.412

Mass

Machine Learning Technique	Training Dataset			Testing Dataset		
	RMSE	R^2	MAE	RMSE	R^2	MAE
Rational quadratic GPR	0.26699	0.97	0.08	0.118	NA	0.01
Ensemble	0.38828	0.93	0.21917	0.00632	NA	0.004
Cubic SVM	0.70689	0.76	0.50732	0.459	NA	0.04
Fine tree	0.45214	0.90	0.27526	0.305	NA	0.3

Note: The table consists of different machine learning techniques in the first column such as the rational quadratic model, ensemble tree, and support vector machines. The 2nd, 3rd, and 4th columns consist of root-mean-square error, R squared, and mean absolute error values for the training dataset, and 5th, 6th, and 7th columns have the same information but for the testing dataset. The table also compares dataset created for uniaxial compression and compression at 45 degrees angle.

total number of observations. Finally, the square root of this value is taken to obtain the RMSE. The mathematical operations are given in Equations (5.1) and (5.2).

$$RMSE = \sqrt{\frac{\sum_{i=1}^{N}(x_i - y_i)^2}{N}}, \tag{5.1}$$

$$MAE = \frac{\sum_{i=1}^{N}|y_i - x_i|}{N}, \tag{5.2}$$

where RMSE = root-mean-square error, MAE = mean absolute error, x_i = true values, y_i = predicted values, N = total number of observations, and i = variable.

5.4   DISCOVERY OF NEW LATTICE UNIT CELLS

In order to predict lattice unit cells with optimal performance, the machine learning models developed in the previous section shall be utilized to estimate the mass and compression stress of several untrained lattice unit cell fingerprints. Using MATLAB, more than 500,000 combinations of untrained lattice unit cells are generated with the "nchoosek" function within the RVE boundary conditions. It can be understood that although a huge dataset of lattice unit cells can be generated, not all the structures have better performance. To identify lattice unit cells with optimal performance, the octet unit cell which is widely considered to have superior structural performance is considered the datum point for comparisons. Once a dataset of several of these untrained combinations is formed, the GPR model is used to predict their mass and compression properties by using the "yfit" function in a time span of 15 minutes or less. The advantage of using a machine learning approach comes into play here as several lattice unit cells can be evaluated for their performance within a few minutes with minimal manual effort and standard computation capacities. After predicting the desired properties, the huge datasets are truncated to only contain lattice unit cells that perform better than the octet lattice unit cell. Though the training dataset generation process requires time and manual effort, once a suitable regression model is selected, the data truncation and identification of optimal lattice unit cells can be rapid. Also, all the possible combinations within set boundary conditions or constraints can be evaluated with minimal manual effort and time consumption. By altering the boundary conditions in the data filtering process, lattice unit cells of any desired properties can be identified. Excel data filtering or MATLAB coding can be used for topology optimization by setting any desired boundary conditions to identify optimal lattice unit cells. For example, topology optimization can be performed by setting certain boundary conditions like low mass, high compression strength, and symmetric or orthotropic truss distribution along the unit cell during the filtering process. In this study, as the octet unit cell is considered as the datum point, a filter is set to extract various fingerprints that have higher relative compression strength (with respect to the overall density) compared to the octet unit cell. Once a dataset of several optimal unit cells is obtained, more filters to extract fingerprints from the new dataset can be set. These filters include much lower mass or higher compression strength in different loading orientations and filters to extract fingerprints that have structural symmetry in the unit cells.

Although classical topology optimization can also be used to design new lattice unit cells, the technique works on the material removal method from a given reference structure. Therefore, although advanced topology optimization techniques can be implemented to design complex structures, the application of this technique can be complex that requires robust coding. Furthermore, classical topology optimization may be limited to a small design space, while machine learning can search optimal structures out of a large design space and thus may predict even better structures than classical topology optimization.

Once the predictions of the mechanical properties for the dataset not used for training are accomplished using the GPR models, they are used to compare with the performance of the octet lattice unit cell evaluated under the same boundary conditions. Following this scheme, a total of 20 optimal designs for symmetric lattice unit cells with superior mechanical properties over octet lattice unit cells are proposed (see Table 5.3).

TABLE 5.3
Selected Optimal Lattice Unit Cells

Unit Cells	Fingerprints	Unit Cells	Fingerptins	Unit Cells	Fingerprints
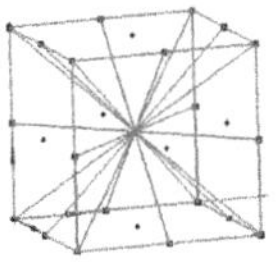	12 18 28	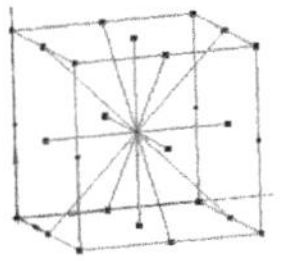	12 28 48	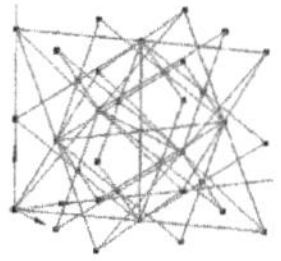	16 45 48
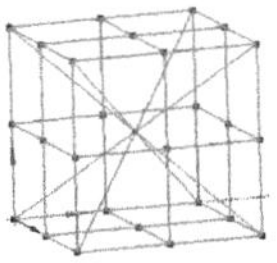	12 18 24	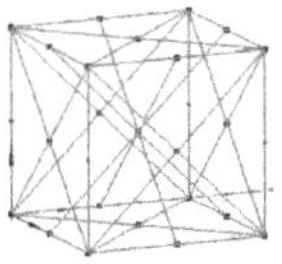	12 16 28	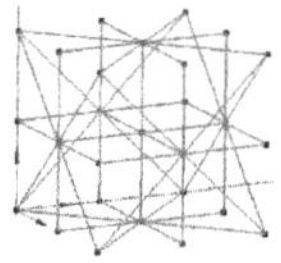	16 24 48
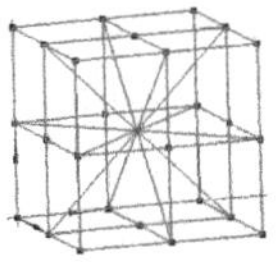	12 24 28	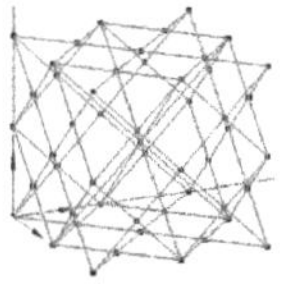	16 25 28	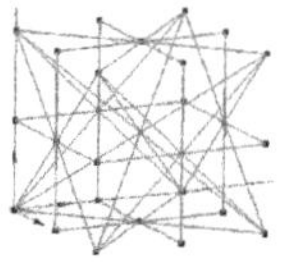	16 24 18
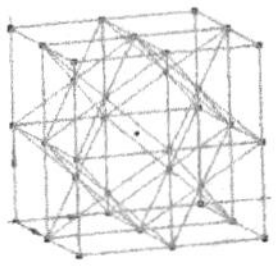	12 24 45	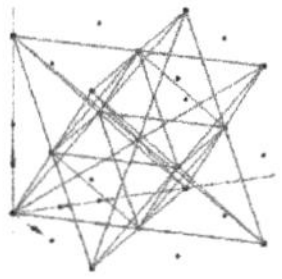	16 18 46	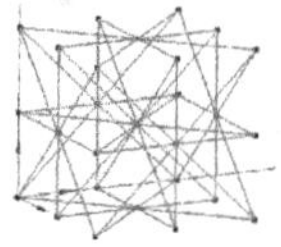	16 24 28
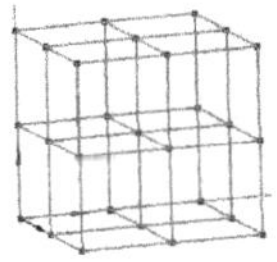	12 24 48	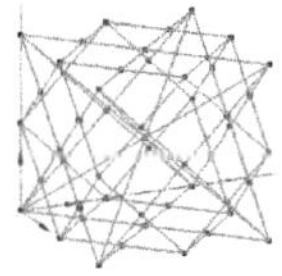	16 18 25	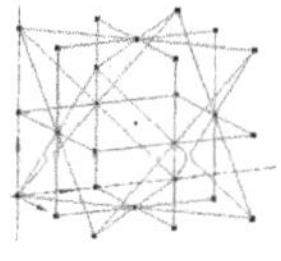	16 24
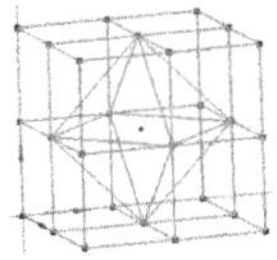	12 24 46	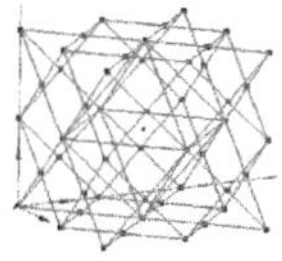	16 25 46	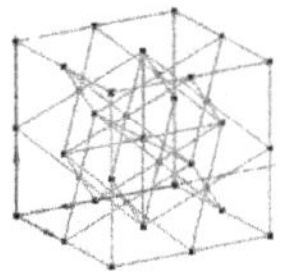	12 48 45
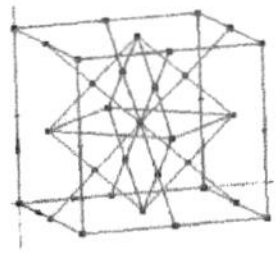	12 846	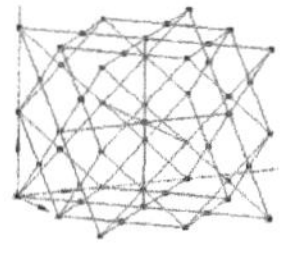	16 25 48		

5.5   NUMERICAL AND EXPERIMENTAL VALIDATIONS FOR NEWLY DISCOVERED LATTICE UNIT CELLS

A portion of these optimal lattice unit cells proposed in Table 5.3 were manufactured through 3D printing and tested under uniaxial compression. For additive manufacturing of the selected optimal lattice unit cells, SolidWorks design tool is used to model and create stereolithographic (STL) files of the three-dimensional structures. These STL files can be read by any type of 3D printer. A professional 3D printer (Pico 2), which uses a vat photopolymerization technique to cure materials, is used for the manufacturing process. All the unit cells are manufactured using the photopolymer. Three samples of each unit cell are manufactured with varying relative densities; see images in Figure 5.5. Once the postprocessing was complete, an MTS machine (ADMET eXpert 2610 Tabletop 5kN Universal Test System) was used to conduct a uniaxial compression test on all the samples. The compression tests were conducted at a speed of 1mm/min, and the load and displacement for each sample were recorded to get the stress–strain curves. ANSYS simulations were conducted on all the samples tested maintaining the same material properties and boundary conditions. Boundary conditions were applied on both the top and bottom surfaces of the unit cells to simulate the compression behavior. The bottom surfaces were fixed in the Z direction, which was in the direction of the applied load, while the top surfaces on which the load was applied were allowed to freely move in the Z direction. One of the other two directions on both the top and bottom surfaces was allowed to move to account for the effect of minor sliding. Mesh convergence was checked on the same structures

FIGURE 5.5 Three-dimensional printed various lattice unit cells.

FIGURE 5.6   Meshing and deformed shape of the lattice unit cells.

used for additive manufacturing which were directly imported into the ANSYS platform for consistency. Figure 5.6 shows the meshing and deformed shapes of the lattice unit cells. All the lattice unit cells failed through brittle fracture, at low strain. Nonlinearity due to imperfections in the 3D-printed lattice unit cells can be observed in Figure 5.7. The sample with fingerprint (12 24 46) performed better compared to the rest of the 3D-printed lattice unit cells in terms of compression strength. The slight variation in the experimental and simulation curves for (12 18 28) unit cells may be due to cured extra polymer resin that was not cleaned properly or left-over support protrusions. Lists of more optimal lattice unit cells with both symmetric and asymmetric structures are presented in Table 5.4 and Table 5.5, respectively.

Once the numerical results are validated by experimental results on a portion of unit cells predicted by machine learning, further simulation analysis is conducted on more optimal lattice unit cells with varying mass to observe the compression and tension stresses induced in the lattice elements under uniaxial compression in different orientations to compare with octet lattice unit cell. From Figure 5.8, it can be observed that compared to the octet lattice structure, the proposed optimal lattice unit cells have lower elemental compressive and tensile stresses with the same mass in normal and angular orientations. More simulation results of other optimal lattice unit cells are presented in Figure 5.9.

Results of simulations for buckling capacities were also compared between the proposed optimal lattice unit cells and octet unit cells as seen in Figure 5.10. It can be seen that the optimal lattice unit cells with fingerprints (12 24 48) and (12 24 46) exhibit 57% and 51% better relative buckling capacities, respectively, while fingerprints (16 25 46) and (16 24) were able to be in par with octet unit cell.

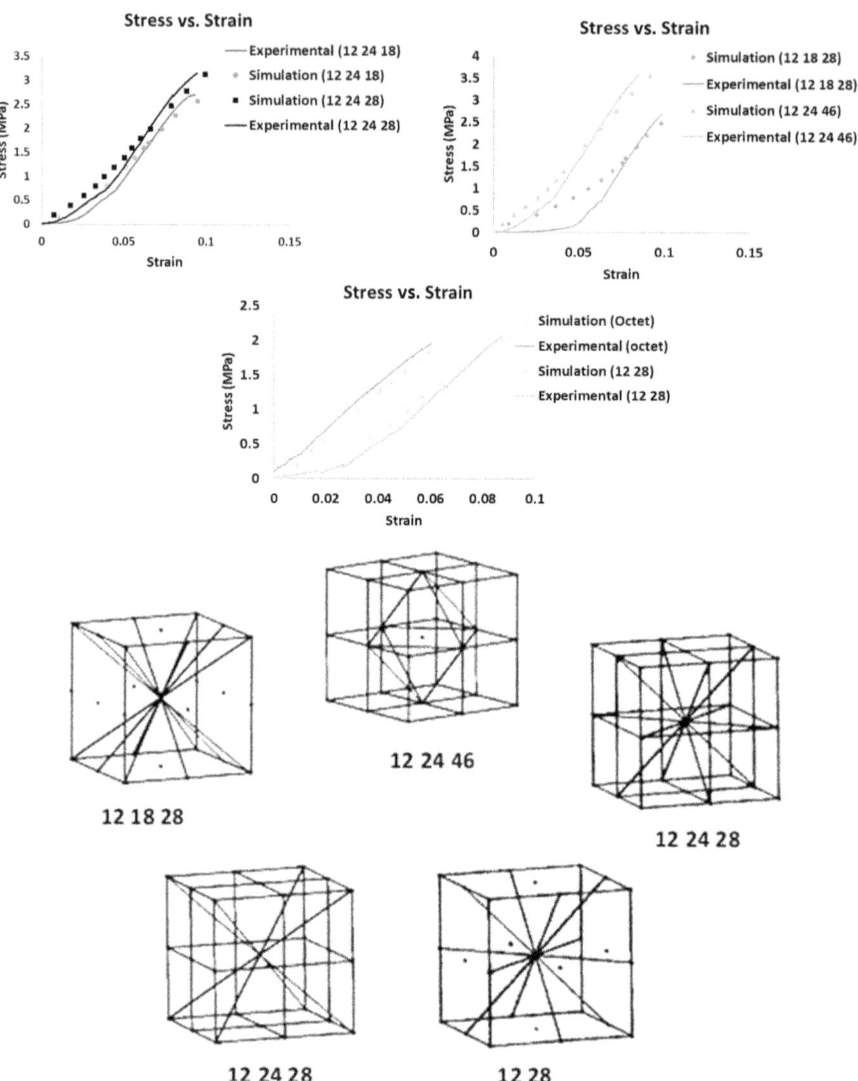

FIGURE 5.7 Experimental and simulation stress–strain curves of various optimal lattice unit cells along with octet truss unit cell (top), and symmetric optimal lattice unit cells with fingerprints (bottom).

To further optimize the buckling performance of the proposed optimal lattice unit cells, the conventional solid rods (elements) of each lattice unit cell can be replaced with biomimetic rods. These biomimetic rods, inspired by the cellular structures of various plant stems, have varying outer shapes and inner porous structures. This porous nature of the biomimetic rods helps in enhancing the buckling capacity of lattice unit cells under compression [62]. Simulation analysis

TABLE 5.4

Complete List of Optimal Symmetric Lattice Unit Cells Proposed through Machine Learning

12 28	16 24	16 25	16 45	12 16 25
12 16 28	12 16 45	12 18 24	12 18 28	12 18 45
12 24 25	12 24 28	12 24 46	12 24 48	12 25 28
12 28 48	12 46 45	12 48 45	16 18 24	16 18 25
16 18 46	16 18 45	16 24 25	16 24 28	16 24 46
16 24 48	16 24 45	16 25 28	16 25 46	16 25 48
16 25 45	16 28 46	16 28 45	16 46 45	16 48 45

TABLE 5.5

Sample List of Optimal Asymmetric Lattice Unit Cells Proposed through Machine Learning

Asymmetric optimal lattice unit cell in the normal direction (unidirectional)

12 13 15 213 1314 1415 1015 910 39 519 1316 1625 1518 1827 911 1121 1922 2225 2526 2627 1920 2021 2124 2427 18 89 813 815 819 821 825 82725 23 35 28 58 46 47 67 46 68 78 45 27

12 13 15 213 1314 1415 1015 910 39 519 1316 1625 1518 1827 911 1121 1922 2225 2526 2627 1920 2021 2124 2427 24 34 37 56 57 26 25 23 35 28 38 58 46 68

12 13 15 213 1314 1415 1015 910 39 519 1316 1625 1518 1827 911 1121 1922 2225 2526 2627 1920 2021 2124 242718 89 813 815 819 821 825 82724 34 37 26 25 23 35 28 38 58 46 47 67 46 68 78 45

12 13 15 213 1314 1415 1015 910 39 519 1316 1625 1518 1827 911 1121 1922 2225 2526 2627 1920 2021 2124 2427 18 89 813 815 819 821 825 82724 34 37 56 57 26 25 23 35 28 38 58 46 47

12 13 15 213 1314 1415 1015 910 39 519 1316 1625 1518 1827 911 1121 1922 2225 2526 2627 1920 2021 2124 2427 24 34 37 56 57 26 25 23 35 28 38 58 46 68 78 45 27

12 13 15 213 1314 1415 1015 910 39 519 1316 1625 1518 1827 911 1121 1922 2225 2526 2627 1920 2021 2124 242718 89 813 815 819 821 825 82725 23 35 46 47 67 46 68 78 45 27 36

12 13 15 213 1314 1415 1015 910 39 519 1316 1625 1518 1827 911 1121 1922 2225 2526 2627 1920 2021 2124 2427 25 23 35 28 46 47 67 46 68 78 45 27 36

12 13 15 213 1314 1415 1015 910 39 519 1316 1625 1518 1827 911 1121 1922 2225 2526 2627 1920 2021 2124 2427 24 34 37 56 57 26 28 67 46 68 78 45 27 36

12 13 15 213 1314 1415 1015 910 39 519 1316 1625 1518 1827 911 1121 1922 2225 2526 2627 1920 2021 2124 2427 24 34 37 56 57 26 25 68 78 45 27 36

12 13 15 213 1314 1415 1015 910 39 519 1316 1625 1518 1827 911 1121 1922 2225 2526 2627 1920 2021 2124 2427 24 34 37 56 57 26 25 23 35 28 38 58 78 45 27 36

16 14 17 625 613 619 912 1227 1215 1221 79 719 721 49 413 415 1317 1517 1725 1727 1923 2123 2325 2327 18 89 813 815 819 821 825 82724 34 37 56 57 35 28 38 58 46 68 78 45 27 36

Asymmetric optimal lattice unit cell in 45° orientation direction (unidirectional)

16 14 17 625 613 619 912 1227 1215 1221 79 719 721 49 413 415 1317 1517 1725 1727 1923 2123 2325 2327 18 89 813 815 819 821 825 82724 34 37 56 57 26 25 23 35 28 45 27 36

16 14 17 625 613 619 912 1227 1215 1221 79 719 721 49 413 415 1317 1517 1725 1727 1923 2123 2325 2327 18 89 813 815 819 821 825 82724 34 37 56 57 26 38 58 46 47 67 46 68 78

(Continued)

TABLE 5.5 *(Continued)*

Sample List of Optimal Asymmetric Lattice Unit Cells Proposed through Machine Learning

16 14 17 625 613 619 912 1227 1215 1221 79 719 721 49 413 415 1317 1517 1725 1727 1923 2123 2325 2327 18 89 813 815 819 821 825 82724 34 37 56 57 26 25 23 35 48 68 36

16 14 17 625 613 619 912 1227 1215 1221 79 719 721 49 413 415 1317 1517 1725 1727 1923 2123 2325 2327 625 613 619 912 1227 1215 1221 79 719 721 49 413 415 1317 1517 1725 1727 1923 2123 2325 2327 56 57 26 25 23 35 46 47 67 45 27 36

16 14 17 625 613 619 912 1227 1215 1221 79 719 721 49 413 415 1317 1517 1725 1727 1923 2123 2325 2327 18 89 813 815 819 821 825 82724 34 26 25 23 35 46 47 67 46 68 78

16 14 17 625 613 619 912 1227 1215 1221 79 719 721 49 413 415 1317 1517 1725 1727 1923 2123 2325 2327 18 89 813 815 819 821 825 82724 34 37 56 57 26 28 38 58 45 27 36

16 14 17 625 613 619 912 1227 1215 1221 79 719 721 49 413 415 1317 1517 1725 1727 1923 2123 2325 2327 18 89 813 815 819 821 825 82724 34 37 56 57 26 25 23 35 28 38 78

16 14 17 625 613 619 912 1227 1215 1221 79 719 721 49 413 415 1317 1517 1725 1727 1923 2123 2325 2327 18 89 813 815 819 821 825 82724 34 37 56 57 26 38 58 46 47 67

16 14 17 625 613 619 912 1227 1215 1221 79 719 721 49 413 415 1317 1517 1725 1727 1923 2123 2325 2327 18 89 813 815 819 821 825 82724 34 37 56 57 26 25 23 35 28 38 58 46 68 78 45

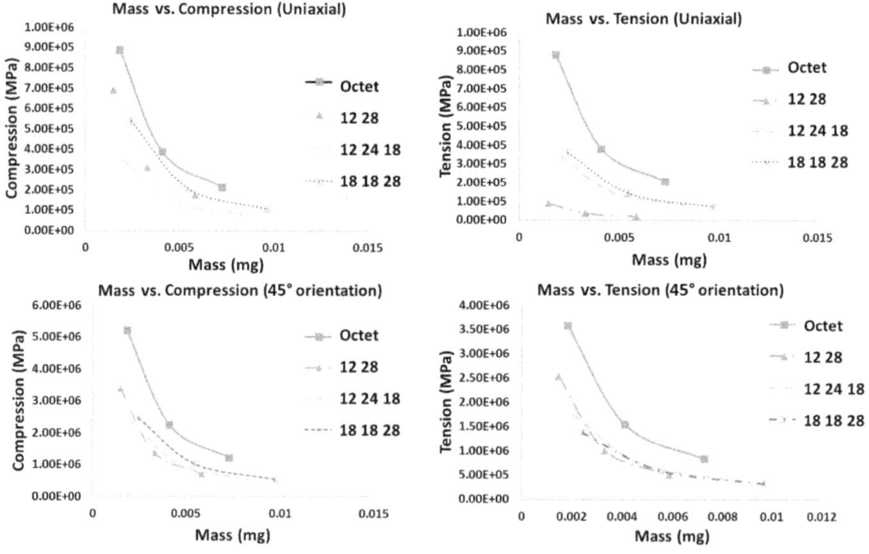

FIGURE 5.8　ANSYS simulation results in comparisons for stresses induced in different optimal lattice unit cells compared with octet lattice unit cells.

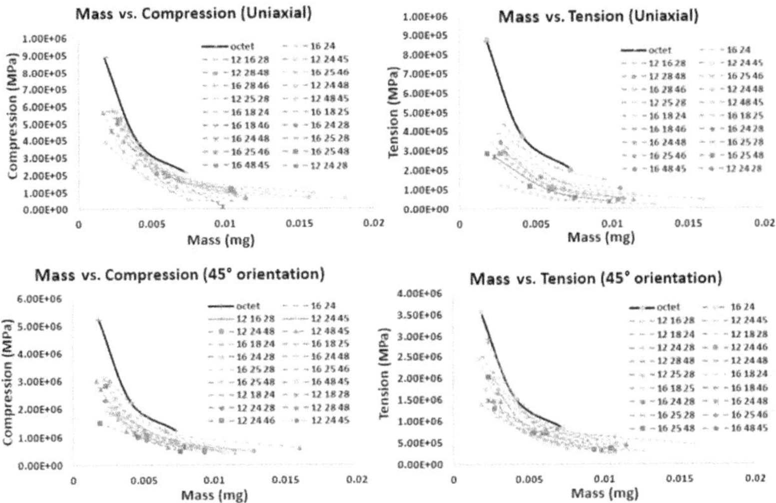

FIGURE 5.9 Several optimal symmetric lattice truss unit cells compared to octet truss under uniaxial compression in axial and angular orientation.

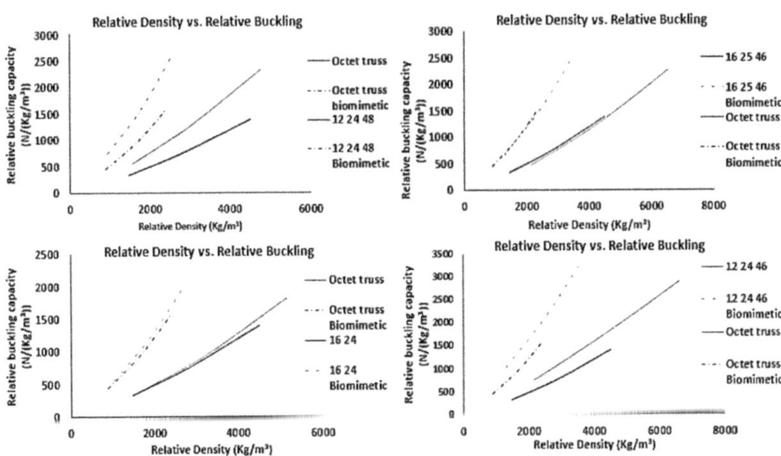

FIGURE 5.10 ANSYS comparisons for relative buckling capacity of various optimal lattice unit cells compared with octet unit cell.

for the buckling capacities of optimal lattice unit cells with biomimetic porous rods can be seen in Figure 5.11. The relative buckling capacity of the optimal lattice unit cells with biomimetic rods is 130–160% higher than that of the optimal lattice unit cells with solid rods. Images of solid designs and simulation deformation shapes and 3D-printed biomimetic rods and lattice unit cells can be found in Figures 5.12 and 5.13, respectively. Finally, example optimal lattice unit cells based on different deformation criteria are listed in Table 5.6.

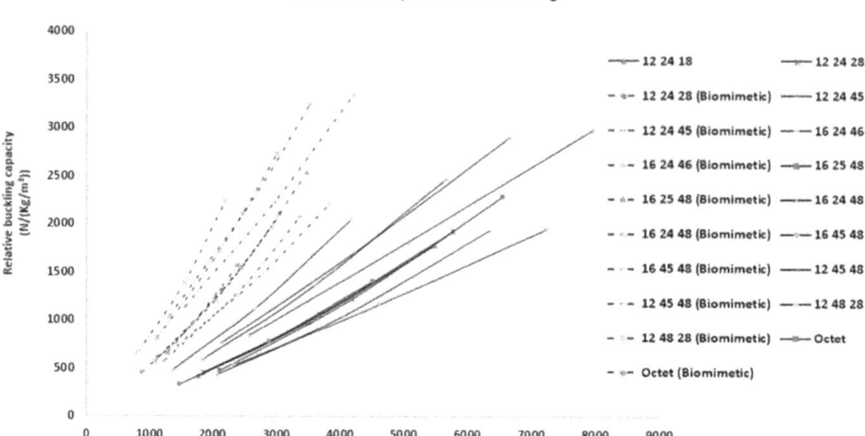

FIGURE 5.11　　Buckling analysis of optimal lattice unit cells with biomimetic rods (dotted curves) and solid cylindrical rods (solid curves).

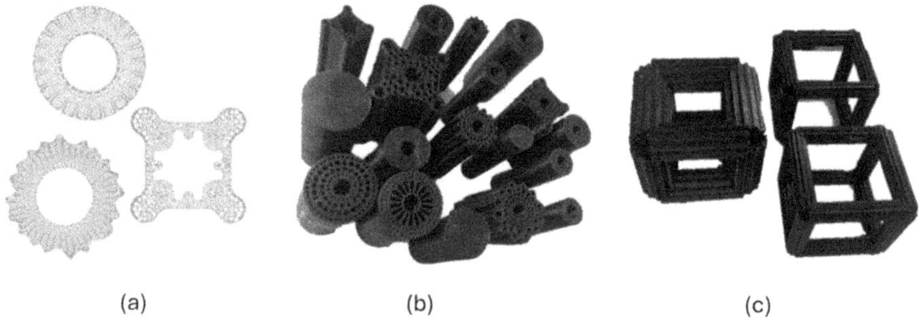

FIGURE 5.12　　(a) Two-dimensional representation of biomimetic rod cross section, (b) 3D printed biomimetic rods, and (c) 3D printed square lattice unit cells with biomimetic rods.

FIGURE 5.13　　Images of optimal lattice unit cells with biomimetic roads using SolidWorks design (gray) and buckling deformation of the corresponding optimal lattice unit cell by ANSYS (colored).

TABLE 5.6

Example Optimal Lattice Unit Cells Based on Different Deformation Criteria

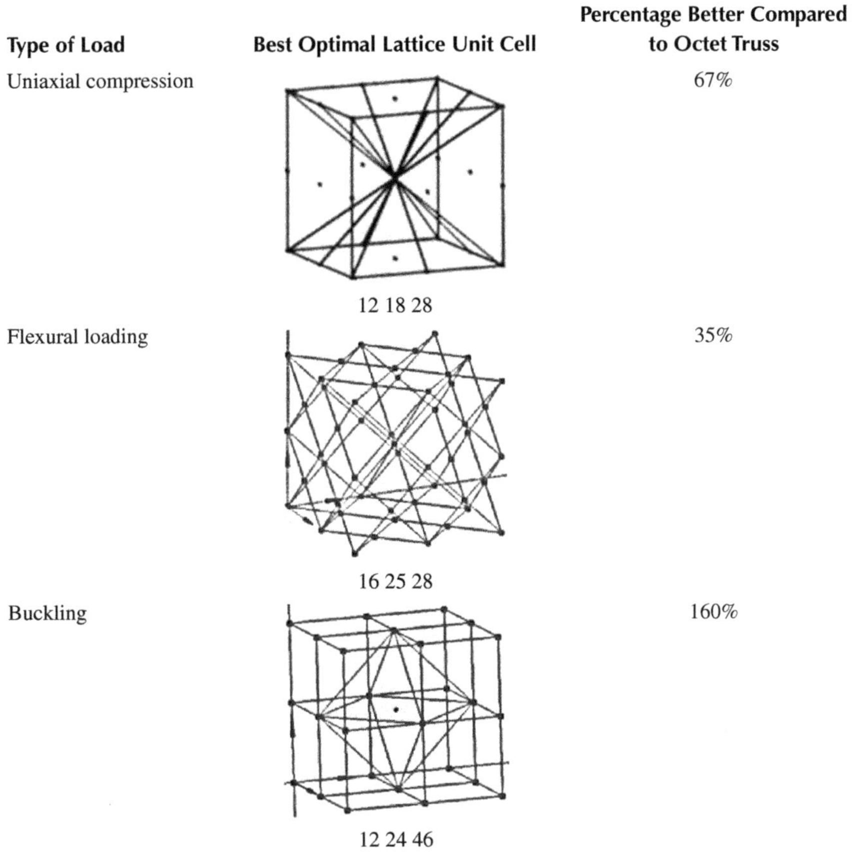

Type of Load	Best Optimal Lattice Unit Cell	Percentage Better Compared to Octet Truss
Uniaxial compression	12 18 28	67%
Flexural loading	16 25 28	35%
Buckling	12 24 46	160%

5.6　THREE-POINT BENDING TEST OF SEVERAL LATTICE-CORED SANDWICH STRUCTURES

The optimal lattice unit cells predicted by machine learning were used to design several sandwich structures with varying densities. Thin sheets with 10% of the thickness of the core are used to laminate the lattice structure constructed by stacking four rows and ten columns of unit cells side by side. Three-point bending tests are performed on all the samples experimentally and compared with ANSYS simulation results. All the tests are performed on samples printed using a commercially available polymer (VeriGuide, tensile strength: 28.5 MPa, and elastic modulus: 1.14 GPa), printed by the Pico 2 printer. After postprocessing the 3D-printed parts, the ADMET eXpert Test System is used to conduct the

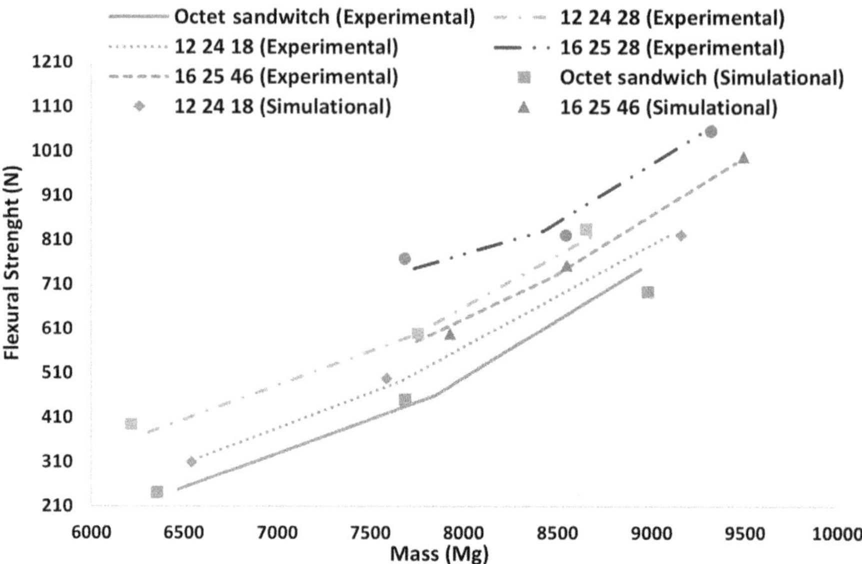

FIGURE 5.14 Experimental comparisons of flexural strength vs mass of various sandwich lattice structures made of the optimal unit cells with octet lattice sandwich structure.

three-point bending tests at a loading rate of 1 mm/min. For the ANSYS simulations, the support rollers are assigned to be made of steel. The contact between the support rods and sandwich structure is frictionless represented by contact interface treatment. Meshing is done using tetrahedral elements, and mesh convergence is tested. Large deflection is set to "ON" to consider geometric nonlinearity. The three-point bending test is performed by applying displacement to the top support rod and fixing the bottom two rollers. It is seen in Figure 5.14 that the ANSYS simulation results are in good agreement with the experimental results. The slight difference can be attributed to the imperfection in the manufacturing and post-processing of the 3D-printed sandwich structures. It can be observed from Figure 5.14 that the flexural strength of sandwich structures made of the optimal symmetric unit cells is higher (13–35%) than the octet lattice sandwich structure. It is observed that sandwich structures with truss elements oriented at an angle (45°) to the sandwich plates perform better compared to sandwich panels made of elements that are perpendicular to the plates and the primary failure modes in the core are joint debonding and struct fracture. Since the entire sandwich structure is additively manufactured as a single part, no debonding of face sheets from the lattice core is observed. The images of the 3D-printed sandwich structures, three-point bending test fixture, and ANSYS simulations can be found in Figure 5.15.

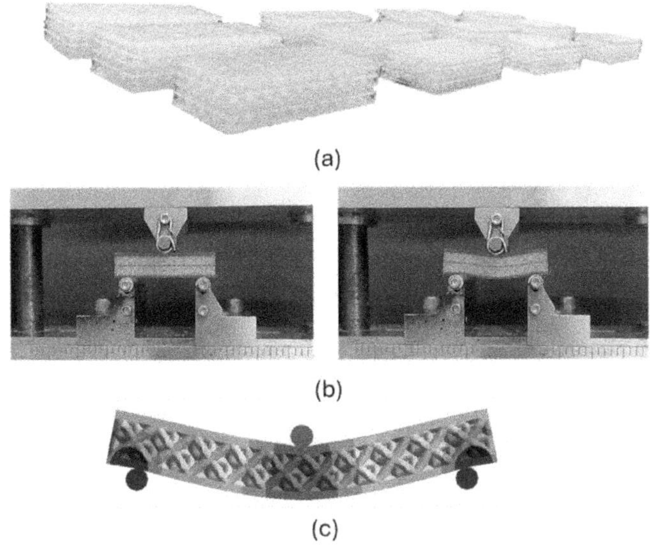

(a)

(b)

(c)

FIGURE 5.15 (a) Several lattice 3D-printed sandwich structures, (b) a sandwich structure with no applied load (left) and just before fracture (right), and (c) ANSYS simulation.

5.7 SAMPLE CODING

```
import numpy as np
from sklearn.gaussian_process import
GaussianProcessRegressor
from sklearn.gaussian_process.kernels import RBF, Matern,
RationalQuadratic
from sklearn.model_selection import train_test_split
from sklearn.metrics import mean_squared_error, r2_score,
mean_absolute_error
# Step 1: Initialize Dataset
# Assume X_train, X_test are the lattice unit cell
fingerprints, and y_train, y_test are the mechanical
properties
# Load your dataset or generate synthetic data
# Step 2: Preprocessing
# Normalize the input features if necessary
# Split the dataset into training and testing sets
X_train, X_test, y_train, y_test = train_test_split(X, y,
test_size=0.2, random_state=42)
# Step 3: Training the GPR Model
# Choose the appropriate kernel function (RBF, Matern,
RationalQuadratic)
kernel = RationalQuadratic(length_scale=1.0, alpha=0.1)
```

```python
# Initialize and train the Gaussian Process Regression model
model = GaussianProcessRegressor(kernel=kernel, random_state=0)
model.fit(X_train, y_train)
# Step 4: Validation
# Evaluate the trained model on the testing set
y_pred = model.predict(X_test)
# Compute evaluation metrics
mse = mean_squared_error(y_test, y_pred)
r2 = r2_score(y_test, y_pred)
mae = mean_absolute_error(y_test, y_pred)
print("Mean Squared Error:", mse)
print("R-squared:", r2)
print("Mean Absolute Error:", mae)
# Step 5: Prediction
# Make predictions on new, unseen data
new_data = np.array([[…]]) # Insert new lattice unit cell
fingerprints here
new_predictions = model.predict(new_data)
print("Predicted mechanical properties:", new_predictions)
# Step 6: Visualization (Optional)
# Visualize the predicted mechanical properties against the
actual values
# Plotting code can be added here using libraries like
matplotlib or seaborn
# Step 7: Iterative Refinement (Optional)
# Fine-tune the model by adjusting hyperparameters or
incorporating additional data
# Repeat steps 3 to 6 as needed
# Step 8: Optimization (Optional)
# Explore advanced techniques like sparse Gaussian processes
for large datasets
# Optimize computational efficiency and predictive
performance
# Step 9: Deployment (Optional)
# Deploy the trained model for practical applications
# Integrate it into software tools or platforms used by
engineers and researchers
# Step 10: Monitoring and Maintenance (Optional)
# Regularly monitor the performance of the deployed model
# Update the model as needed to accommodate changes in data
distribution or requirements.
```

Please note that this code is a template and needs to be customized according to the specific dataset and requirements. To use the code, please ensure that the necessary libraries are installed on the computer (scikit-learn, numpy, and matplotlib for visualization) before running the code. Additionally, replace the placeholder "…" with the actual data for new lattice unit cell fingerprints when making predictions.

5.8 SUMMARY

A lattice structure is a framework created by stacking lattice unit cells together in various configurations. The performance of the entire lattice structure hinges on the design of the individual lattice unit cell. Research in this area has focused on the design, fabrication, and evaluation of lightweight lattice architectures. Depending on the number of struts and joints in a unit cell, it can be categorized as stretching- or bending-dominated. Stretching-dominated structures perform better in terms of strength and stiffness compared to foam and bending-dominated lattice structures. Various lattice unit cell designs have been proposed with superior performance in structural, thermal, impact, vibrational, optical, and acoustic domains. Examples include the octet lattice structure, gyroid structures, and hollow micro-truss lattice structures.

Fabrication techniques and structural performance of lattice-cored sandwich structures have been explored extensively. Techniques such as selective laser melting (SLM) and SLA have been used to manufacture lattice cores with sandwich plates made of carbon fiber-reinforced face sheets.

Machine learning techniques, such as SVM and GPR, have been utilized to predict the mechanical properties of lattice structures and identify optimal designs.

Experimental validation of optimized lattice unit cells through 3D printing and uniaxial compression testing has been conducted. Results show that the predicted mechanical properties closely match experimental observations. Moreover, sandwich structures made of optimal lattice unit cells exhibit higher flexural strength compared to traditional lattice structures. When the solid rods are replaced by the porous biomimetic rods discovered in Chapter 4, the buckling resistance of the lattice unit cells can be further improved.

REFERENCES

1. Gibson LJ, Ashby MF. Cellular Solids: Structure and Properties. Cambridge University Press, (1999).
2. Saba N, Jawaid M, Alothman OY, Paridah MT, Hassan A. Lightweight Composites from Natural Fibers: Development and Applications. Springer, (2016).
3. Wu W, Jiang H, Shi Y. Additive manufacturing of gyroid structures with superior impact absorption capability. Scientific Reports, 7:1–9, (2017).
4. Yang C, Li Y, Sun G, Li L, Zhang J. Hollow micro-truss lattice structures for enhanced energy absorption. International Journal of Impact Engineering, 127:130–141, (2019).
5. Adhikari S, Pankaj P, Chakraborty S. Mechanical and damping properties of pyramid truss sandwich panels. Composite Structures, 224:111103, (2019).
6. Tobia M, Martin L, Bill L, Zhan M, Qia O, Faruque M. SLM lattice structures: Properties, performance, applications and challenges. Materials & Design, 183:108137, (2019).
7. Deshpande VS, Fleck NA, Ashby MF. Effective properties of the octet-truss lattice material. Journal of Mechanics and Physics of Solids, 49:1747–1769, (2001).

8. Radek V, Dane K, David P. Impact resistance of different types of lattice structures manufactured by SLM. Science Journal, 6:1579–1585, (2016).

9. Evans AG, He MY, Deshpande VS, Hutchinson JW, Jacobsen AJ, Barvosa-Carter W. Concepts for enhanced energy absorption using hollow micro-lattices. International Journal of Impact Engineering, 37:947–959, (2010).

10. Yang JS, Ma L, Rüdiger S. Hybrid lightweight composite pyramidal truss sandwich panels with high damping and stiffness efficiency. Composites Structures, 148:85–96, (2016).

11. Zhang X, Chen H, Li M, Liu X, Li J. Bioinspired kagome sandwich panels with titanium core. Composites Science and Technology, 170:179–187, (2019).

12. Yan C, Hao L, Hussein A, Bubb S. Design and testing of bio-inspired sandwich structures with hierarchical honeycomb cores. Composite Structures, 108:636–645, (2014).

13. Lee HJ, Kang KJ. Bending response of graded lattice core sandwich structures. Composite Structures, 211:134–145, (2019).

14. Lee JY, Kim HJ, Hong SW, Choi WJ, Choi JH. Development of tetrahedral core sandwich structures using CFRP laminates. Composite Structures, 97:349–356, (2013).

15. Mao Y, Guan Z, Liu Y. Topology optimization of lattice structures with design-dependent loading. Journal of Applied Mechanics, 82:031006, (2015).

16. Tancogne-Dejean T, Mohr D. Elastically-isotropic truss lattice materials of reduced plastic anisotropy. International Journal of Solids and Structures, 138:24–39, (2018).

17. Ang XJ, Li YJ. Topology optimization of a novel cuttlebone-like lattice for improved energy absorption. International Journal of Impact Engineering, 132:103289, (2019).

18. Messner MC. Optimal lattice-structured materials. Journal of the Mechanics and Physics of Solids, 96:162–183, (2016).

19. Yang S, Cui Y, Chen Y. Multiscale fuzzy optimization of octet-truss lattice structure for crashworthiness design. Composite Structures, 210:889–900, (2019).

20. Nasrullah M, Hayat N, Liao WH, Shahzad M. A comparative study of the mechanical behavior of additively manufactured regular lattices for crashworthiness applications. International Journal of Impact Engineering, 117:85–99, (2018).

21. Watts KC. Topology optimization of maximally stiff multiscale structures. International Journal of Solids and Structures, 134:17–32, (2018).

22. Wang J, Chen W. Numerical homogenization for periodic lattice structures. Journal of the Mechanics and Physics of Solids, 60:2030–2048, (2012).

23. Pan C, Han YF, Lu JP. Design and optimization of lattice structures: A review. Applied Science-BASEL, 10:6374, (2020).

24. Liu H, Long LC. Equivalent homogenization design method for stretching-bending hybrid lattice structures. Journal of Mechanical Science and Technology, 37:4169–4178, (2023).

25. Yin HF, Zhang WZ, Zhu LC, Meng FB, Liu JE, Wen GL. Review on lattice structures for energy absorption properties. Composite Structures, 304:116397, (2023).

26. Vasiliev VV, Barynin VA, Razin AF. Anisogrid composite lattice structures – Development and aerospace applications. Composite Structures, 94:1117–1127, (2012).

27. Hunt CJ, Morabito F, Grace C, Zhao Y, Woods BKS. A review of composite lattice structures. Composite Structures, 284:115120, (2022).

28. Kechagias S, Karunaseelan KJ, Oosterbeek RN, Jeffers JRT. The coupled effect of aspect ratio and strut micro-deformation mode on the mechanical properties of lattice structures. Mechanics of Materials, 191:104944, (2024).

29. Dong Y, Chen KJ, Liu H, Li J, Liang ZH, Kan QH. Adjustable mechanical performances of 4D-printed shape memory lattice structures. Composite Structures, 334:117971, (2024).
30. Wagner MA, Lumpe TS, Chen T, Shea K. Programmable, active lattice structures: Unifying stretch-dominated and bending-dominated topologies. Extreme Mechanics Letters, 29:100461, (2019).
31. Khaderi SN, Deshpande VS, Fleck NA. The stiffness and strength of the gyroid lattice. International Journal of Solids and Structures, 51:3866–3877, (2014).
32. Xiao LJ, Xu X, Feng GZ, Li S, Song WD, Jiang ZX. Compressive performance and energy absorption of additively manufactured metallic hybrid lattice structures. International Journal of Mechanical Sciences, 219:107093, (2022).
33. Zhang MY, Yang ZY, Lu ZX, Liao BH, He XF. Effective elastic properties and initial yield surfaces of two 3D lattice structures. International Journal of Mechanical Sciences, 138:146–158, (2018).
34. Vafaeefar M, Moerman KM, Vaughan TJ. Experimental and computational analysis of energy absorption characteristics of three biomimetic lattice structures under compression. Journal of the Mechanical Behavior of Biomedical Materials, 151:106328, (2024).
35. Bhat C, Kumar A, Lin SC, Jeng JY. Design, fabrication, and properties evaluation of novel nested lattice structures. Additive Manufacturing, 68:103510, (2023).
36. Torquato S. Random Heterogeneous Materials: Microstructure and Macroscopic Properties. Springer Science & Business Media, (2002).
37. Gurson AL. Continuum theory of ductile rupture by void nucleation and growth: Part i-yield criteria and flow rules for porous ductile media. Journal of Engineering Materials and Technology, 99:2–15, (1977).
38. Deshpande VS, Ashby MF, Fleck NA. Foam topology: Bending versus stretching dominated architectures. Acta Materialia, 49:1035–1040, (2001).
39. Kollmann HT, Abueidda DW, Koric S, Guleryuz E, Sobh NA. Deep learning for topology optimization of 2D metamaterials. Materials & Design, 196:109098, (2020).
40. Wilt JK, Yang C, Gu GX. Accelerating auxetic metamaterial design with deep learning. Advanced Engineering Materials, 22:1901266, (2020).
41. Chang YF, Wang H, Dong QX. Machine learning-based inverse design of auxetic metamaterial with zero Poisson's ratio. Materials Today Communications, 30:103186, (2022).
42. Wang MH, Sun S, Zhang TY. Machine learning accelerated design of auxetic structures. Materials & Design, 234:112334, (2023).
43. Peng XL, Xu BX. Data-driven inverse design of composite triangular lattice structures. International Journal of Mechanical Sciences, 265:108900, (2024).
44. Liu S, Acar P. Generative adversarial networks for inverse design of two-dimensional spinodoid metamaterials. AIAA Journal, 62:2433–2442, (2024).
45. Kumar S, Tan SH, Zheng L, Kochmann DM. Inverse-designed spinodoid metamaterials. NPJ Computational Materials, 6:73, (2020).
46. Zheng L, Kumar S, Kochmann DM. Data-driven topology optimization of spinodoid metamaterials with seamlessly tunable anisotropy. Current Methods in Applied Mechanics and Engineering, 383:113894, (2021).
47. Ma W, Cheng F, Liu YM. Deep-learning-enabled on-demand design of chiral metamaterials. ACS Nano, 12:6326–6334, (2018).
48. Ashalley E, Acheampong K, Besteiro LV, Yu P, Neogi A, Govorov AO, Wang ZM. Multitask deep-learning-based design of chiral plasmonic metamaterials. Photonics Research, 8:1213–1225, (2020).

49. Gu LL, Liu HZ, Wei ZC, Wu RH, Guo JP. Optimized design of plasma metamaterial absorber based on machine learning. Photonics, 10:874, (2023).

50. Qiu YH, Chen SX, Hou ZY, Wang JJ, Shen J, Li CY. Chiral metasurface for near-field imaging and far-field holography based on deep learning. Micromachines, 14:789, (2023).

51. Aru M, Ghanshyam P, Tran DH, Turab L, Rami R. Machine learning strategy for accelerated design of polymer dielectrics. Scientific Reports, 10:20952, (2016).

52. Wu S, Kondo Y. Machine-learning-assisted discovery of polymers with high thermal conductivity using a molecular design algorithm. Computational Materials, 5:66, (2019).

53. Yan C, Feng X, Konlan J, Mensah P, Li G. Overcome the barrier: Designing novel thermally robust shape memory vitrimers by establishing a new machine learning framework. Physical Chemistry Chemical Physics, 25:30049–30065, (2023).

54. Yan C, Lin X, Feng X, Yang H, Mensah P, Li G. Advancing flame retardant prediction: A self-enforcing machine learning approach for small datasets. Applied Physics Letters, 122:251902, (2023).

55. Yan C, Li G. The rise of machine learning in polymer discovery. Advanced Intelligent Systems, 5:2200243, (2023).

56. Yan C, Feng X, Li G. From drug molecules to thermoset shape memory polymer: A machine learning approach. ACS Applied Materials & Interfaces, 13:60508–60521, (2021).

57. Yan C, Feng X, Wick C, Peters A, Li G. Machine learning assisted discovery of new thermoset shape memory polymers based on a small training dataset. Polymer, 214:123351, (2021).

58. Cao YF, Wu W, Zhang HL, Pan JM. Prediction of the elastic modulus of self-compacting concrete based on SVM. Trans Tech Publications, 357:1023–6, (2013).

59. Li A, Challapalli A, Li G. 4D printing of recyclable lightweight architectures using high recovery stress shape memory polymer. Scientific Reports, 9:7621, (2019).

60. Li A, Fan J, Li G. Recyclable thermoset shape memory polymer with high stress and energy output via facile UV-curing. Journal of Materials Chemistry A, 6:11479–11487, (2018).

61. Rasmussen CE. Gaussian processes in machine learning. Advanced Lectures on Machine Learning, 3176, (2013).

62. Challapalli A, Li G. 3D printable biomimetic rod with superior buckling resistance designed by machine learning. Scientific Reports, 10:20716, (2020).

6 Inverse Machine Learning Using Generative Adversarial Networks

6.1 INTRODUCTION

In the preceding chapters, we have shown how machine learning techniques can be applied to predict structural properties and suggest better biomimetic rods with superior buckling resistance and optimal lattice unit cells compared to their natural counterparts [1–7]. We have proposed these optimal rods based on plant stems and animal quills and used forward machine learning to extract hidden features in the biomimetic rods and expedite the computation process [6]. We have also proposed symmetric optimal lattice unit cells using forward machine learning, which involved predicting the structural properties of a huge data set of lattice unit cells whose properties were unknown and filtering the data set to select the best lattice unit cells [7].

However, this forward machine learning technique has limitations. The data filtering process is semi-optimal and involves a lot of manual effort, which is time-consuming and may miss out on some of the optimal designs. Furthermore, it is impossible to achieve the purpose of "structures by design," which involves finding the optimal structures given the desired structural properties and a set of constraints such as mass, volume, and strength. Therefore, a better machine learning technique is needed to handle large datasets with minimal time consumption and manual effort and to predict targeted optimal designs based on a set of desired design constraints.

To realize "structures by design," which is similar to "materials by design," the answer persists in inverse design [8–11]. In this chapter, we explore an inverse machine learning framework using generative adversarial networks (GANs) [12, 13]. Unlike forward machine learning, which predicts the desired properties of a chemical or physical structure, the inverse machine learning framework predicts an optimal structure given the desired properties. Inverse design using two deep neural networks has led to faster and more accurate results compared to other numerical techniques [14]. GANs have been used to design photonic crystals, hypothetical inorganic materials, and meta-surfaces [15].

However, the application of inverse design in structural design optimization based on mechanical properties has not been explored. Therefore, in this chapter,

DOI: 10.1201/9781003400165-6

we discuss the first inverse machine learning framework to identify, predict, and optimize targeted lattice unit cell and cellular unit cell designs that can be used to manufacture high-performance structurers such as core in sandwich structures. To create the inverse design framework, GANs are used to generate potential lattice unit cells and thin-walled cellular unit cells, and the rest of the inverse design framework consists of the forward regression model and boundary conditions or desired properties. The optimal lattice unit cells and cellular unit cells predicted using the inverse design frameworks are then used to design sandwich structures and evaluated with numerical and experimental methods.

The predicted lattice unit cells exhibit compression strength of 40–120% higher than octet unit cells with the same overall volume under uniaxial compression. The predicted thin-walled cellular unit cells, on the other hand, exhibit a significant improvement of 300–800% in energy absorption in terms of normalized energy and a 10–50% increase in normalized natural frequency compared to the honeycomb unit cell. With minor adjustments in the initial and boundary conditions, these frameworks can be easily implemented in designing various structures such as plate lattice structures with given properties [16]. The implementation of GANs and machine learning regression models to optimize the structural performance of metamaterials has the potential to revolutionize the field of material science and engineering.

6.2 GANs

GANs are a type of machine learning model that can generate new data samples that are similar to a given dataset. They were first introduced by Ian Goodfellow and his colleagues in 2014 [17] and have since gained widespread attention and applications in various fields [18–43], from biomedical to engineering, from language to graph, and from materials to fault diagnoses. The basic idea behind GANs is to train two neural networks simultaneously: a generator and a discriminator. The generator takes random noise as input and generates new data samples, while the discriminator tries to distinguish between the generated samples and the real data. The two networks are trained in a minimax game, where the generator tries to fool the discriminator by generating samples that are indistinguishable from the real data, while the discriminator tries to correctly identify the real data from the generated samples.

The training process is iterative, with the generator and discriminator networks being updated alternatively. The generator tries to improve its output by generating samples that are more similar to the real data, while the discriminator improves its ability to distinguish between the real and generated samples. The ultimate goal of this process is to train a generator that can generate realistic samples that are similar to the real data. One of the key advantages of GANs is their ability to generate highly realistic and diverse samples, especially when compared to other generative models like variational autoencoders (VAEs) [44].

GANs have become a popular research area in the field of deep learning due to their ability to generate synthetic data that is indistinguishable from real data.

GANs have been applied to a wide range of fields, including image and video synthesis, data augmentation, anomaly detection, style transfer, and speech synthesis. In image and video synthesis, GANs have been used to generate high-quality images and videos with a high degree of realism. For example, StyleGAN has been used to generate high-quality images of faces that look almost indistinguishable from real faces [45]. GANs have also been used to generate synthetic medical images that can be used to train deep learning models for medical image analysis [46]. In anomaly detection, GANs are trained on normal data and used to identify samples that do not fit the normal distribution. This approach has been used for applications such as fraud detection [47] and intrusion detection in computer networks [48]. In style transfer, GANs can transfer the style of one image to another while preserving the content of the original image. This has been used for image-to-image translation and artistic style transfer. For example, the CycleGAN model has been used to translate images of horses to zebras and *vice versa* [49]. Additionally, GANs have been used for video-to-video translation. For instance, in the Deepfake project, GANs were used to generate realistic videos of famous personalities [50]. In speech synthesis, GANs have been used to generate high-quality speech from text, which has applications in text-to-speech systems and virtual assistants. For example, the MelGAN model has been used to generate high-quality speech [51]. Since our first paper published in 2021 on using GANs to discover new lattice unit cells [52], the use of GANs in the design and optimization of metamaterials has been growing [16, 53–59].

In conclusion, GANs have shown great potential in a wide range of fields and are likely to be applied to even more innovative applications in the future. We have provided some seminal papers on the use of GANs in various areas of studies and most recently published papers [18–43]. Readers can refer to these publications for the broader scope of applications of GANs. A sample code using GANs to generate new lattice unit cells will be provided in Section 6.3.3.

6.3 INVERSE DESIGN FRAMEWORK FOR OPTIMIZATION OF 2D AND 3D LIGHTWEIGHT STRUCTURES

In this section, we will discuss two inverse design frameworks based on GANs: one for lattice unit cells and the other for thin-walled cellular unit cells.

6.3.1 Inverse Design Framework for Optimizing Load-Carrying Capacity of Lattice Unit Cells

The study employs a framework for the inverse design of lattice unit cells, which is illustrated in Figure 6.1. The framework includes a training dataset, a discriminator, a generator, and other components. The training dataset is utilized to train the discriminator, which distinguishes between real and fake data generated by the generator. The real data then undergoes a process that involves constraints by the initial and boundary conditions and forward regression model to obtain a new

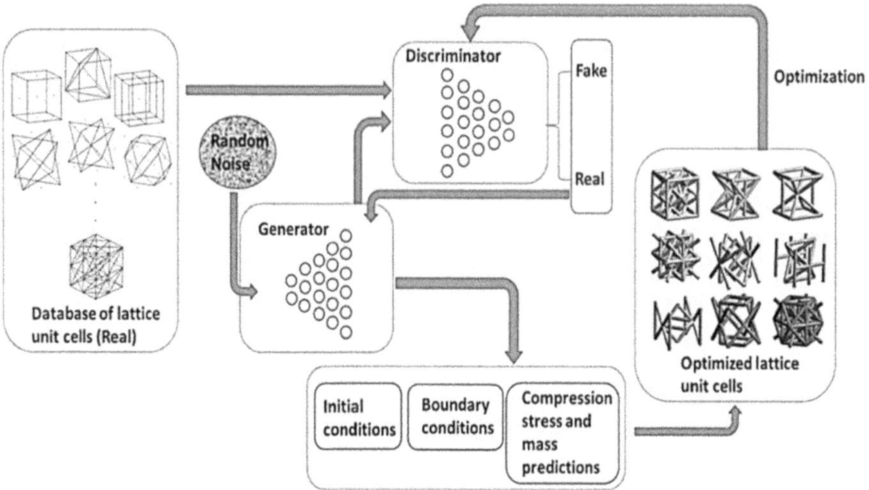

FIGURE 6.1 The inverse design framework for lattice unit cells. The figure shows a representative workflow of 3D lattice unit cells that are fed into the discriminator of a GAN network. The newly generated unit cells are then fed to the code with boundary conditions and finally show the process of extracting optimized unit cells. It later shows the loop of continuously optimizing these unit cells by feeding them back to the GAN system.

set of lattice unit cells with better properties. This new dataset is used to further optimize the lattice unit cells by iteratively training the GAN network and updating the discriminator with the latest datasets.

The framework aims to achieve inverse design by inputting the desired properties for a lattice unit cell, such as low mass, high compression strength, or symmetric truss distribution, and outputting a set of optimal lattice unit cells in the form of fingerprints that fall within the fixed input conditions. Thus, the desired outcome of inverse design, which is predicting optimal lattice structures as output with given properties as input, can be achieved through the framework once the GAN network and forward design regression models are trained.

The inverse design framework presented in this chapter incorporates forward regression analysis as an integral component. The framework involves using the fingerprints of lattice unit cells as input data and developing forward regression models to predict mechanical properties such as mass and compression strength for untrained lattice unit cell designs. However, due to the limited input variables of mass and compression strength, it is not feasible to predict new fingerprints given the mechanical properties as inputs. This is because the desired output would be a large vector with up to hundreds of variables, and the dataset size and structural orientation of the unit cells make it difficult to backpropagate a forward regression model.

To address this limitation, inverse machine learning techniques such as GANs are employed in this chapter. The GANs system in this chapter is trained using the 1500 lattice unit cells in the training database, which are created in Chapter 5. The forward design model is also obtained in Chapter 5. The generator keeps generating new fingerprints until the discriminator cannot differentiate between the fake and real data. The output of the GANs system is a set of unique fingerprints that closely resemble those in the training dataset.

However, the GANs can only generate fingerprints that have similar features to those in the training dataset but are not identical. Thus, the new fingerprints generated by the GANs system should be validated against several criteria, including optimal compression strength, mass, structural authenticity, and whether it is bending- or stretching-dominated. To achieve this, the newly generated dataset is subjected to the constraints of several boundary conditions, and the mass and compression strength are evaluated quickly using the forward regression model generated in Chapter 5. Maxwell's criterion is used to differentiate between bending- and stretching-dominated structures [60]. Moreover, to ensure structural authenticity, a condition is set where at least one end of each truss element must be connected to another element. This eliminates fingerprints with truss elements that lack joints.

Recently, some researchers have extended Maxwell's criterion to 3D printed structures that have elastic, rigid, or frozen joints [61–63]. In this chapter, we used Maxwell's criterion as a rough guide to help us select bending- or stretching-dominated unit cells.

$$M = b - 2j + 3, \text{ 2D structures (frames)}, \tag{6.1}$$

$$M = b - 3j + 6, \text{ 3D structures}, \tag{6.2}$$

where b is the number of truss members and j is the number of frictionless joints. If $M \geq 0$, the structure is stretching-dominated, and if $M < 0$, the structure is bending-dominated.

6.3.2 INVERSE DESIGN FRAMEWORK FOR OPTIMIZING LOAD-CARRYING CAPACITY WITH HIGHER NATURAL FREQUENCY OF THIN-WALLED UNIT CELLS

While lightweight structures such as lattice-cored sandwich have been widely used in load-bearing engineering structures such as fuselage, wing of aircraft, wind turbine blade, ship hull, bridge deck, offshore oil platform, etc., one concern with these lightweight structures persists in their resonance to dynamic load and vulnerability to impact load. Increasing the natural frequency to avoid resonance and enhancing energy absorption are viable ways to improve the performance of these lightweight structures. To this end, biomimetic design has been a driving force for discovering optimal cellular structures.

Biomimetic cellular structures inspired from nature such as honeycombs, plant stems, luffa sponges, trabecular bones, muscles, beetle wings, etc. are widely studied for their excellent specific stiffness, strength, and energy absorption properties

[64–72]. Hexagon-shaped cellular structures inspired from honeycombs have been extensively studied and optimized for their superior energy absorption capacities [67]. Different plant stems like bamboo, rice, and square stems have been mimicked to design cylindrical rods with buckling strength seven times higher than solid and hollow cylinders [68]. The hierarchical bio-cellular structure of luffa sponge with micro- and macropores is mimicked to manufacture foam cylinders reinforced by stiff thin-walled carbon fiber-reinforced polymer (CFRP) tubes with good energy absorption properties [69]. The hierarchical inner structures of tabular bones and muscles are taken as inspiration to design energy-absorbing and impact-resistant tubular sections that exhibited a 176% increase in energy absorption for the third-order hierarchy compared to the first-order hierarchy [70, 71]. The highly efficient energy-absorbent properties of beetle electra provide inspiration to design trabecular honeycomb structures that are five times better than conventional quadrilateral tubes used in the crash box beams of modern devices and vehicles [72]. Frequency optimization of macroscopic structures is studied to be an important criterion to avoid destructive response [73]. Natural frequency is optimized by 40% to aid structures subjected to dynamic loading [74].

Similar to lattice structures, thin-walled cellular structures such as honeycomb have been widely used in engineering structures, for example, honeycomb-cored sandwich structures as lightweight load-bearing structures. Therefore, discovering new thin-walled cellular unit cells is not only of research interest but also of practical value. Upon identifying the most superior regression models, an inverse design framework is formulated to predict optimal cellular unit cells in comparison to honeycomb unit cells and forecast unit cells with desired structural features. This inverse design framework, which is shown in Figure 6.2, is constructed

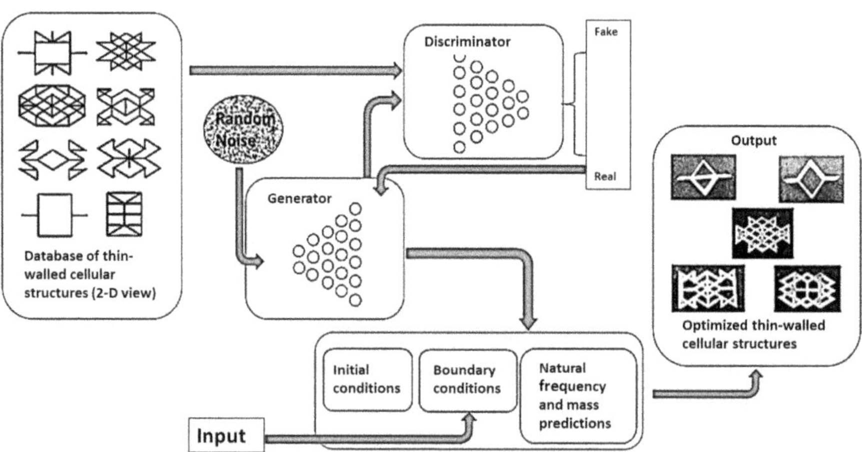

FIGURE 6.2   Inverse design framework for thin-walled unit cells. The framework shows the workflow of feeding a database of thin-walled unit cells to the discriminator of a GAN system. The GAN system then loops into a set of initial and boundary conditions that filter to produce optimal thin-walled unit cell designs.

by amalgamating GANs with forward regression and other predefined boundary conditions. The inverse design framework incorporates GANs, forward regression models, initial conditions, and boundary conditions, where the boundary conditions, such as required mass, maximum load, or natural frequency, can be set to generate new cellular unit cells with desired structural properties. In this study, to train the GAN system, a subset of 300 cellular unit cells that exhibited better natural frequency than honeycomb unit cells were extracted from the initial training dataset and fed to the discriminator of the GAN system as shown in Figure 6.2. The discriminator was iteratively trained with the generator until the generator produced new fingerprints that were highly similar but not identical to the subset of cellular unit cells fed to the discriminator. Hence, new, untrained fingerprints were generated using GANs, and the cellular unit cell properties, such as mass and natural frequency, were predicted using the forward regression models.

To forecast optimal cellular unit cells that outperform biomimetic cellular structures, honeycomb unit cell structural properties are established as boundary conditions, alongside the regression models. Novel optimal cellular unit cell fingerprints that possess higher natural frequencies without sacrificing mass and strength compared to the honeycomb unit cell are generated using the inverse design framework. The input to this framework consists of the desired properties of a cellular unit cell, in the form of boundary conditions, and the output comprises a set of innovative cellular unit cells, as depicted in Figure 6.2. The proposed models' natural frequencies are simulated and compared in the subsequent sections.

The framework proposed for the inverse design of optimal cellular unit cells consists of several steps. Step 1 involves training a GAN system that generates novel cellular unit cell fingerprints. This is achieved by feeding a subset of 300 cellular unit cells that exhibit superior natural frequencies compared to the honeycomb unit cell to the discriminator of the GAN system. The discriminator iteratively trains with the generator until the generator produces new fingerprints that are remarkably similar but not identical to the subset of cellular unit cells fed to the discriminator. This process results in the generation of untrained fingerprints that can be used for the subsequent steps.

In step 2 of the framework, desired properties such as high natural frequency and low mass are fed as inputs to the boundary conditions. These properties function as design constraints for generating optimal cellular unit cells.

In step 3, the newly generated fingerprints from the GAN system are passed through the boundary conditions and forward regression model for mass and natural frequency predictions. The forward regression model is developed using the best regression models identified in the earlier stages of the framework. The predictions are then used to determine whether the generated fingerprints satisfy the boundary conditions set in step 2.

Finally, in step 4, optimal cellular unit cell fingerprints that satisfy the boundary conditions are produced. These optimal cellular unit cells are expected to perform better than biomimetic cellular structures, such as the honeycomb unit cell. Simulation comparisons for natural frequencies of the proposed models are presented in the subsequent sections, demonstrating the effectiveness of the proposed framework. It is noted that while we use natural frequency as an example, this framework also works for using other mechanical/physical properties as the inputs.

6.3.3 Sample GAN Code for Generating Novel Fingerprints Using Python Scripting

With the necessary Python package, the code can be used to define a generator, discriminator, and training variables and finally convert the output into 1s and 0s.

```python
# Training Parameters
learning_rate = 0.0002
batch_size = 107
epochs = 14

# Network Parameters
image_dim = 20
gen_hidd_dim = 10
disc_hidd_dim = 10
z_noise_dim = 5
def xavier_init(shape):
return tf.random_normal(shape = shape, stddev=1./
tf.sqrt(shape[0]/2.0))

#initializing weights and bias for Discriminator and
Generator Neural Networks
weights = {
"disc_H": tf.Variable(xavier_init([image_dim,
disc_hidd_dim])),
"disc_final": tf.Variable(xavier_init([disc_hidd_dim,1])),
"gen_H": tf.Variable(xavier_init([z_noise_dim,
gen_hidd_dim])),
"gen_final": tf.Variable(xavier_init([gen_hidd_dim,
image_dim]))
}
bias = {
"disc_H": tf.Variable(xavier_init([disc_hidd_dim])),
"disc_final": tf.Variable(xavier_init([1])),
"gen_H": tf.Variable(xavier_init([gen_hidd_dim])),
"gen_final": tf.Variable(xavier_init([image_dim]))
}

#defining the Discriminator
def Discriminator(x):
hidden_layer = tf.nn.relu(tf.add(tf.matmul
(x, weights["disc_H"]),
```

```python
bias["disc_H"]))
final_layer = (tf.add(tf.matmul(hidden_layer,
weights["disc_final"]), bias["disc_final"]))
disc_output = tf.nn.sigmoid(final_layer)
return final_layer, disc_output

# Defining the Generator NW
def Generator(x):
hidden_layer = tf.nn.relu(tf.add(tf.matmul(x,
weights["gen_H"]),
bias["gen_H"]))
final_layer = (tf.add(tf.matmul(hidden_layer,
weights["gen_final"]),
bias["gen_final"]))
gen_output = tf.nn.sigmoid(final_layer)
return gen_output
#define placeholders for external input
z_input = tf.placeholder(tf.float32, shape = [None, z_noise_
dim], name
= "input_noise")
x_input = tf.placeholder(tf.float32, shape = [None, image_
dim], name =
"real_noise")
#Building the Generator NW
with tf.name_scope("Generator") as scope:
output_Gen = Generator(z_input) #G(z)

# Building the Disc NW
with tf.name_scope("Discriminator") as scope:
real_output1_disc, real_output_disc = Discriminator(x_input)
#implements D(x)
fake_output1_disc, fake_output_disc = Discriminator(output_
Gen) #
implements D(G(x))

#first kind of loss
with tf.name_scope("Discriminator_Loss") as scope:
Discriminator_Loss=-tf.reduce_mean(tf.log(real_output_
disc+0.0001)+tf.log(1.fake_output_disc+0.0001))

# LF= log(D(x))+log(1-D(G(z)));
with tf.name_scope("Genetator_Loss") as scope:
Generator_Loss = -tf.reduce_mean(tf.log(fake_output_disc+
0.0001)) #
due to max log(D(G(x)))
#LF= log(1-D(G(z))) -> -log(D(G(z)));
# T-board summary
Disc_loss_total = tf.summary.scalar("Disc_Total_loss",
Discriminator_Loss)
```

```python
Gen_loss_total = tf.summary.scalar("Gen_loss",
Generator_Loss)
# load the features into the numpy array
dat = np.loadtxt('Paper4FingerPrints.txt')

Generator_var = [weights["gen_H"], weights["gen_final"],
bias["gen_H"], bias["gen_final"]]
Discriminator_var = [weights["disc_H"],
weights["disc_final"],
bias["disc_H"], bias["disc_final"]]

#Define the optimizer
with tf.name_scope("Optimizer_Discriminator") as scope:
Discriminator_optimize=tf.train.AdamOptimizer(learning_
rate=learning_rate).minimize(Discriminator_Loss, var_list
=Discriminator_var)
with tf.name_scope("Optimizer_Generator") as scope:
Generator_optimize = tf.train.AdamOptimizer(learning_rate =
learning_rate).minimize(Generator_Loss, var_list =
Generator_var)

# Initialize the variables
init = tf.global_variables_initializer()
sess = tf.Session()
sess.run(init)
writer = tf.summary.FileWriter("./log", sess.graph)
i = 0
for epoch in range(epochs):

#dividing into batches
x_batch = np.array(dat[i:i+batch_size])
# x_batch, _ = mnist.train.next_batch(batch_size)
#Generate noise to feed Discriminator
z_noise = np.random.uniform(-1.,1.,size = [batch_size,
z_noise_dim])

#Discriminator
_, Disc_loss_epoch = sess.run([Discriminator_optimize,
Discriminator_Loss], feed_dict = {x_input:x_batch,
z_input:z_noise})

#Generator
_, Gen_loss_epoch = sess.run([Generator_optimize,
Generator_Loss],
feed_dict = {z_input:z_noise})

#Running the Discriminator summary
summary_Disc_loss = sess.run(Disc_loss_total, feed_dict =
{x_input:x_batch, z_input:z_noise})
```

```
# Adding the Discriminator summary
writer.add_summary(summary_Disc_loss, epoch)

#Running the Generator summary
summary_Gen_loss = sess.run(Gen_loss_total, feed_dict =
{z_input:z_noise})

# Adding the Generator summary
writer.add_summary(summary_Gen_loss, epoch)

#convert from features from values between 0s and 1s to only
0s and 1s
def conversion(gen_vector):
for i in range(20):
if gen_vector[i] < 0.7:
gen_vector[i] = 0
else:
gen_vector[i] = 1
return gen_vector
z_noise = np.random.uniform(-1.,1., size = [batch_size,
z_noise_dim]).
```

6.4   DATA GENERATION, FINGERPRINTING, AND FORWARD DESIGN MODEL OF CELLULAR UNIT CELLS

In Chapter 5, we have presented the details on how to develop a forward design model for lattice unit cells. As shown in Figures 6.1 and 6.2, the forward design mode is an integrated component in the inverse design framework. Therefore, in this section, we will focus on developing the forward design model for thin-walled cellular unit cells.

6.4.1   DATASET GENERATION AND FINGERPRINTING

Data generation and fingerprinting are crucial parts of any study involving machine learning applications. In order to train a regression model to predict the mass, compression load, and natural frequency of various cellular unit cells, a training dataset needs to be formed. In this study, the structures in focus are thin-wall cellular unit cell structures that exhibit higher natural frequencies compared to honeycomb and other biomimetic structures. The height of all the unit cells and their wall thickness are constant for simplicity in generating new designs. A representative volume element (RVE) as shown in Figure 6.3(a) with 9 vertices is formed in a 2D format. Using vertices of the RVE, several lines can be drawn by connecting any two neighboring vertices. Combining a few lines that connect different neighboring vertices will form a quarter of the unit cell. Now, by mirroring this combination of lines on horizontal and vertical axes, a 2D image can

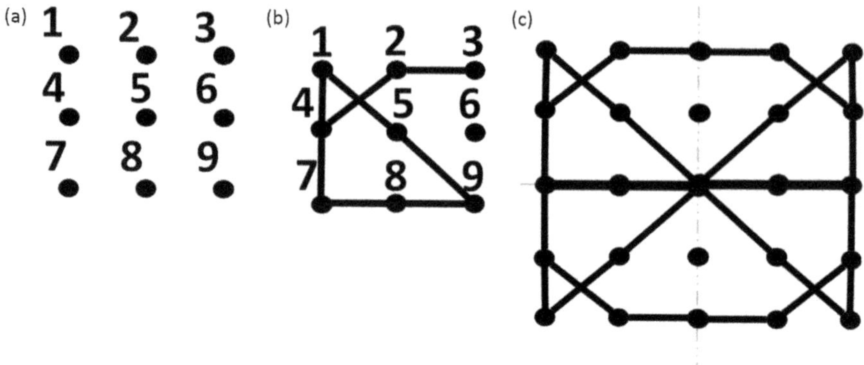

FIGURE 6.3 RVE for thin-walled unit cell data generation. (a) Representative volume element (RVE), (b) a quarter of the cellular unit cell, and (c) a full cellular unit cell formed by mirroring the RVE. The RVE in the figure consists of nine points in two dimensions, forming a 3 by 3 matrix. These points are numbered from 1 to 9. To form an entire unit cell, the 3 by 3 RVE is mirrored in both horizontal and vertical directions to form a 2D unit cell with a matrix of 25 by 25 points. There are nine common points along the horizontal and vertical mirroring lines. As a result, only $4 \times 9 - 9 = 25$ independent points within the entire unit cell.

be formed. By assigning thickness to the lines and extruding them in the third dimension, a full 3D unit cell is formed.

Fingerprinting is the process of converting the unit cell designs into a consistent machine-readable format. Since all the designs are formed by mirroring the RVE on horizontal and vertical axes and have the same wall thickness and height, these features do not contribute to the machine learning process and can be omitted from the fingerprints. This reduces the fingerprinting process to the initial RVE which only consists of thin lines connecting the nine vertices in the 3 by 3 matrix. For fingerprinting, all the vertices are named from 1 to 9, and each line formed by connecting the neighboring vertices is named after the two vertices it connects. Now, the fingerprint of a single unit cell is formed by combining the names of all the lines forming that unit cell. For example, in Figure 6.3(b), the lines connecting vertices 1 and 4, 1 and 5, and 2 and 4 are named as 14, 15, and 24, respectively. The rest of the lines are named in a similar manner, and the combination of all the line names (14 15 23 24 47 59 78 89) will be the fingerprint of the single unit cell in Figure 6.3(b). While designing the unit cell for numerical simulations, the fingerprint design from the RVE is mirrored into the horizontal and vertical axes to form a complete unit cell as stated earlier. Using this fingerprinting process, a design can be easily inferred from a fingerprint, or a design can be easily converted into a fingerprint or digital representation. For forward regression and inverse design, these fingerprints are further converted into a vector of 1s and 0s for more accurate machine learning predictions. This is done by assigning a 20-vector space (12 14 15 23 24 25 26 35 36 45 47 48 56 57 58 59 68 69 78 89) for all the possible lines of the RVE in a vector and by placing "1" in the vector if a particular fingerprint consists of that line and by placing "0s" in the rest of the spaces. It is noted that if all the 9 vertices are used to form an RVE, the maximum number of lines that can be created are the 20 lines shown above in the 20-element vector. For example, the final fingerprint of the design in Figure 6.3(b) will be of the form (0 1 1 1 1 0 0 0 0 0 1 0 0 0 0 1 0 0 1 1). Consistent boundary conditions should be set for generating all the data points, and the same technique should be adopted to fingerprint all the designs for logical forward regression and inverse design.

In this study, we created 2,000 cellular unit cells, and ANSYS software package was used to analyze the mechanical and physical properties such as compressive strength and natural frequency. This dataset was used to train a forward design model, i.e., a model that correlates the fingerprints to the mechanical and physical properties of the cellular unit cells.

6.4.2 REGRESSION OR FORWARD DESIGN MODEL TRAINING FOR THIN-WALLED UNIT CELLS

After the data generation and fingerprinting process, a training dataset is constructed comprising 2,000 fingerprints. The aim was to train machine learning models for forward regression, which would predict the mass, load, and natural frequency of various cellular unit cells. To accomplish this, ANSYS workbench tools were employed to estimate the mass, load, and first natural frequency of each fingerprint under uniaxial compression from the training dataset. Natural frequency plays a pivotal role in dynamic structural loading, and for simplicity, only the first natural frequency of the cellular unit cells was considered. This is because the rest of the natural frequencies follow a similar trend, which simplifies the regression process. The theoretical background of the first natural frequencies will be presented in the last section of this chapter.

To evaluate the accuracy of various regression models in predicting the mass, maximum load, and natural frequency of new cellular unit cells, we used the MATLAB regression analysis tool. The regression models tested included ensemble trees, Gaussian process regression (GPR), and support vector machines (SVMs). The quadratic SVM model provided the most accurate mass predictions, with a root-mean-square error (RMSE) of 0.0048 kg. On the other hand, the GPR models were the most effective in predicting the maximum load and natural frequency of cellular unit cells, with RMSE values of 0.16628 N and 0.8031 Hz, respectively. The performance of these models was further evaluated through prediction vs. response plots, as shown in Figure 6.4. The nearly symmetrical scattering of the points along the diagonal line suggests that these models are dependable and accurate in predicting the mass, maximum load, and natural frequency of cellular unit cells.

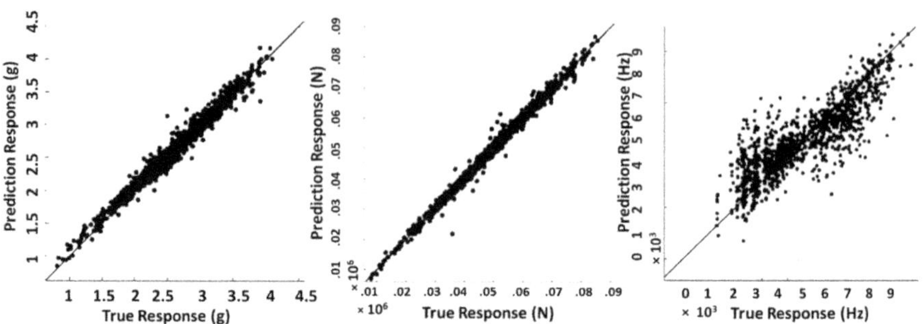

FIGURE 6.4 True vs. prediction responses of mass (quadratic SVM), load (cubic GPR), and natural frequency (rational quadratic GPR) from left to right, respectively. The figure depicts the evenly rough distribution of predictions vs. the true response, showing this is a suitable model for the given data.

6.5 NUMERICAL AND EXPERIMENTAL VALIDATION

6.5.1 VALIDATION FOR LATTICE UNIT CELLS

To ensure that the training dataset created by finite element modeling is reliable, the modeling results need to be validated by experiments. In this comprehensive investigation, the lattice unit cells postulated and assessed within the confines of this study are meticulously crafted utilizing three-dimensional computer-aided design (CAD) software, specifically SolidWorks. The subsequent stage involves subjecting these CAD designs to a rigorous examination of their compression behavior, employing ANSYS Workbench for both linear and nonlinear analyses. The transition from SolidWorks to ANSYS is facilitated through the conversion of CAD designs into an industry-standard, vendor-neutral file format known as IGES.

The material chosen as the foundation for all numerical and experimental validations is Veri Guide, a commercially available photopolymerizable and 3D printable resin boasting a tensile strength of 28.5 MPa and an elastic modulus of 1.14 GPa. The simulation structures undergo meshing using tetrahedron elements, with a meticulous assessment for large deflections to ensure compatibility with nonlinear analysis. Mesh convergence is meticulously verified across various sizes, ultimately adopting an adaptive sizing strategy with a resolution order of 4, equating to approximately 900,000 elements. This judicious choice strikes a delicate balance between expeditious simulation speed and convergence, as elucidated in Figure 6.5 showcasing the meshed lattice unit cells and their deformed configurations under uniaxial compression loading.

For the experimental facet, stereolithographic (STL) files representing the three-dimensional lattice structures are generated through SolidWorks, providing compatibility with a myriad of 3D printers. The manufacturing process transpires using a state-of-the-art 3D printer, specifically Pico 2, leveraging vat photopolymerization techniques for material curing. The printer exclusively employs Veri Guide as the commercial polymer, with all unit cells uniformly adhering to a

FIGURE 6.5 Mesh and deformation of few lattice unit cells. The figure shows the meshing of three different unit cells and then the deformation contour of the unit cells under compression. The unit cells compression shows stress concentration in vertical trusses to be compression and horizontal trusses to be tension.

volumetric specification of $20 \times 20 \times 20$ mm. Subsequent to postprocessing, the mass of each unit cell undergoes meticulous measurement using a dual-range XS105 balance. The subsequent phase involves subjecting the manufactured lattice structures to uniaxial compression tests conducted at a controlled speed of 1 mm/min, with load and displacement data meticulously recorded to generate comprehensive load vs. displacement curves.

Because of the brittleness of the polymer used, all the units succumb to brittle fracture at low strain levels. The ensuing analysis, as depicted in Figure 6.6, reveals a commendable alignment between simulation results and empirical data. Any marginal deviations observed between the two datasets are attributed to potential imperfections in the 3D printed components or the inadvertent retention of uncured resin and support materials during the postprocessing phase.

Gibson and Ashby [64, 65] and Gibson [66] conceived a model elucidating the modulus and strength of cellular structures and predicated upon the linear elastic characteristics at diverse relative densities. This model is expressed by the following equation:

$$E/E_s = c_1\left(\rho/\rho_s\right)^n \; ; \; \sigma_e/E_s = c_2\left(\rho/\rho_s\right)^n \; ; \; \sigma_p/\sigma_y = c_3(\rho/\rho_s)^n, \qquad (6.3)$$

where E is Young's modulus of the porous material, E_s is Young's modulus of the cell wall material (solid), ρ is the density of the porous material, ρ_s is the density

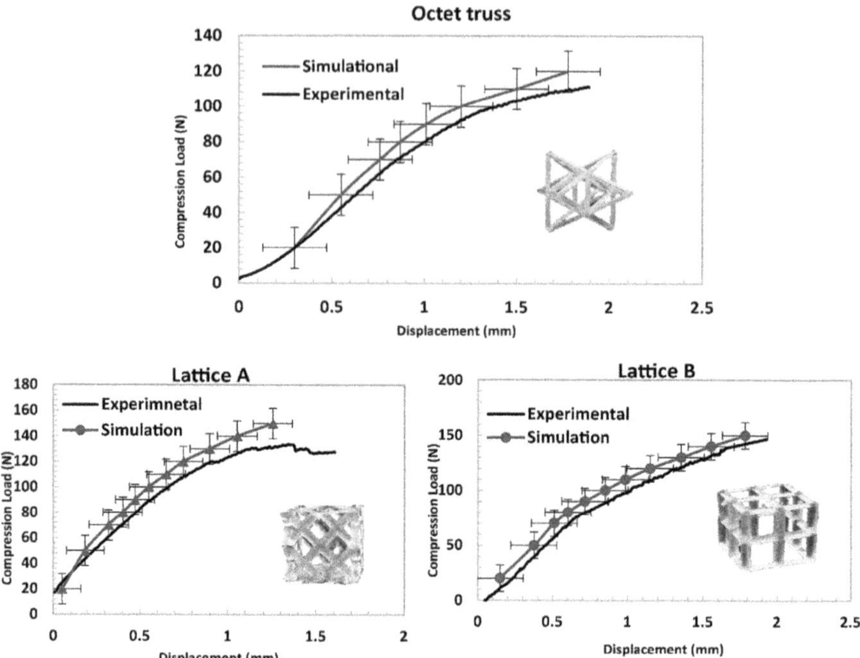

FIGURE 6.6 Experimental vs. simulation comparison under uniaxial compression. Three-unit cells are considered: an octet unit cell, an optimal unit cell A, and an optimal unit cell B. The results show that the curves of simulation and experiments are in good agreement with each other, suggesting that the finite element modeling used in creating the training dataset is reliable.

of the cell wall material (solid), σ_e is the elastic collapse stress of the porous material (cell wall buckles), σ_p is the plastic collapse stress of the porous material (cell wall yields), σ_y is the yield strength of the cell wall material (solid), n is the scaling factor, and c_1, c_2, and c_3 are material-dependent constants. Based on the literature, $n = 1.5 \sim 3$, depending on if the cell is closed or open [64–66]. It is clear from Equation 6.3 that for ultralow-density porous materials, the mechanical properties degrade significantly. For example, if the relative density is 10% and $n = 3$, Young's modulus and collapse stress become 0.1% of their original values. Therefore, the grand challenge in porous materials is how to achieve high strength and stiffness with minimal weight penalty.

The c_1 values ascribed to the octet, lattice A, and lattice B unit cells, as delineated in Figure 6.6, are quantified at 0.09, 0.1, and 0.125, respectively. The n value is taken as 2. These values are derived from a comprehensive analysis of experimental results conducted across various densities. Figure 6.7 provides a nuanced perspective on the relative modulus of the octet unit cell vis-à-vis its dependence on the relative density of the unit cell. Intriguingly, the discerning observation emerges that the relative modulus of the octet unit cell exhibits a comparatively diminished

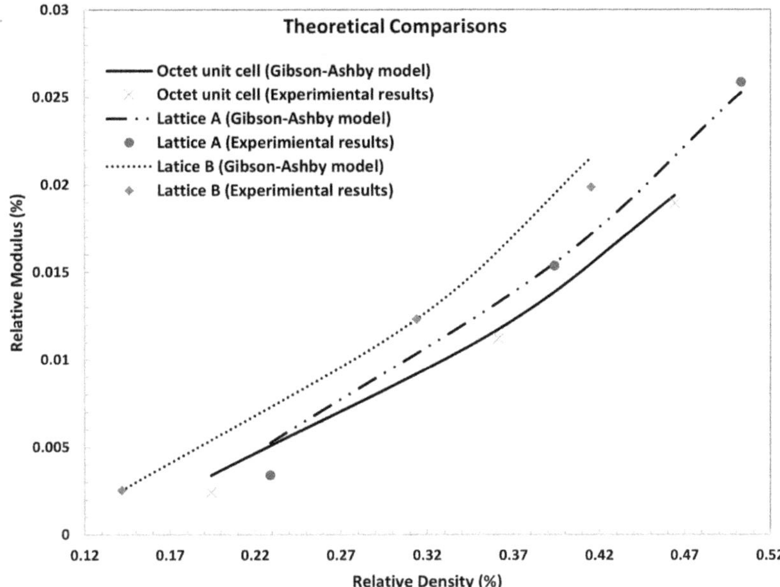

FIGURE 6.7 Gibson–Ashby model predictions comparisons of the relative modulus of lattice unit cells. The figure shows the relative modulus vs. relative density curves for octet unit cell and lattice unit cell a and b calculated using Gibson–Ashby models.

reliance on variations in relative density. In simpler terms, when confronted with equivalent relative densities, unit cells A and B manifest elevated modulus ratios. Translating this empirical insight into a practical realm, wherein the same material is employed for the unit cells octet, A, and B, yields a noteworthy implication. Specifically, unit cells A and B boast superior stiffness when juxtaposed with the octet unit cell. This heightened stiffness in unit cells A and B is a salient and coveted characteristic, especially in the context of load-carrying structures. In essence, the model posited by Gibson and Ashby, enriched by the empirical determination of values, serves as a predictive tool for understanding the nuanced interplay between relative density and modulus in cellular structures. The quantified values offer not only a mathematical framework but also practical insights into the design considerations for load-bearing elements, shedding light on the advantageous features embodied by specific lattice unit cells. This intersection of empirical observation and theoretical modeling stands as a testament to the sophisticated understanding attained in the realm of cellular materials and their mechanical properties.

6.5.2 Validation for Thin-Walled Cellular Unit Cells Using Uniaxial Compression

A comprehensive comparison between the finite element modeling and experimental testing was conducted to validate the model, which is the basis for trusting

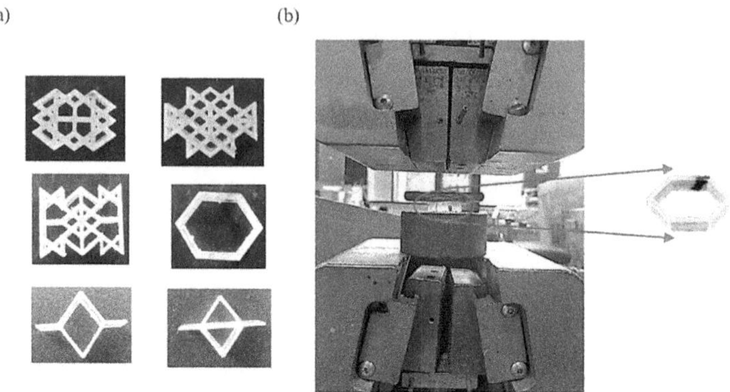

FIGURE 6.8   Additively manufactured (a) cellular unit cells and (b) uniaxial compression.

the training dataset created by the finite element modeling. Uniaxial compression tests on several thin-walled cellular unit cells were conducted.

A few fingerprints of cellular unit cell structures based on their superior performance in natural frequency compared to honeycomb structured unit cells are selected to observe their uniaxial compression behavior. These structures are designed into 3D models using a CAD design software (SolidWorks). All the 3D models were converted into STL files and are 3D printed using an STL 3D printer and photopolymer procured from Formlabs as shown in Figure 6.8(a). The clear photopolymer has a density of 1.16 g/cm^3 and a compressive strength of 225 MPa. After postprocessing, a Q-TEST 150 machine is used to conduct uniaxial compression tests on all the samples. The maximum load before failure for each sample is recorded for comparisons and validations. ANSYS design modeler and simulation software were used for the numerical analysis. The 3D models from SolidWorks were converted into XML format and imported to the ANSYS design modeler for pre-processing. The bulk material properties like the density, Poisson's ratio, and stress–strain curves from uniaxial compression were uploaded directly into the software. These bulk material properties are obtained from the uniaxial compression test of 3D printed cylinders (12.7 mm × 12.7 mm × 25.4 mm) by following the ASTM D695-15 standard for 3D printable polymers; see a typical test result in Figure 6.9. Constant printing orientation for calibrating the material properties and manufacturing the cellular structures are followed. A mesh convergence test is conducted for the design to obtain consistent results without affecting the computational time too much; see Figure 6.10. Figure 6.11 shows the comparison between the ANSYS simulation and experimental results of several cellular unit cells under uniaxial compression. It is seen that the modeling results and experimental results agree with each other, suggesting that the ANSYS simulation is reliable, and the dataset created by the ANSYS simulation can be used for training the machine learning model.

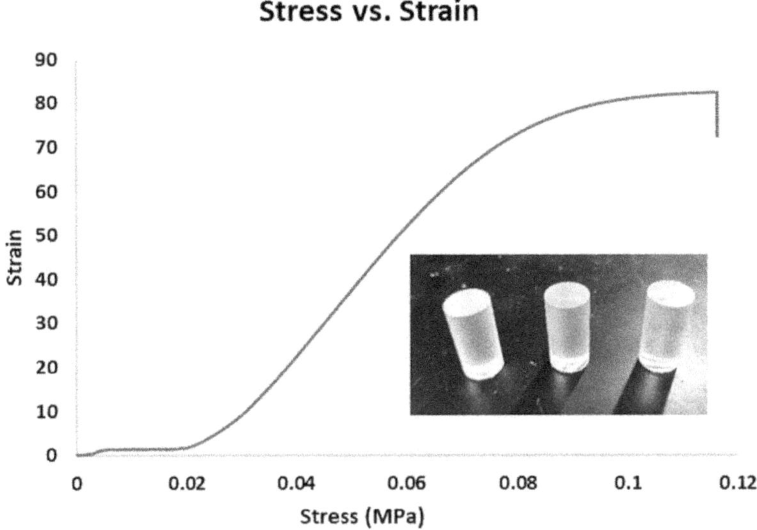

FIGURE 6.9 Typical uniaxial compression test results for 3D printed cylinders.

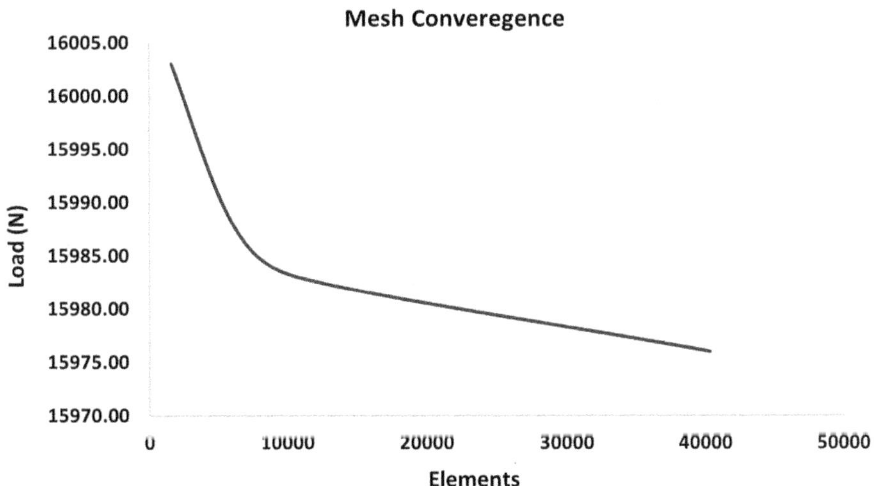

FIGURE 6.10 Convergence analysis of the ANSYS finite element modeling.

6.6 RESULTS AND DISCUSSIONS

In the previous sections, we have established the inverse design framework, created training datasets, established forward a design regression model, and validated the finite element modeling results with experiments. In this section, we will report the new lattice unit cells and cellular unit cells discovered by the inverse design framework and discuss the simulation and test results.

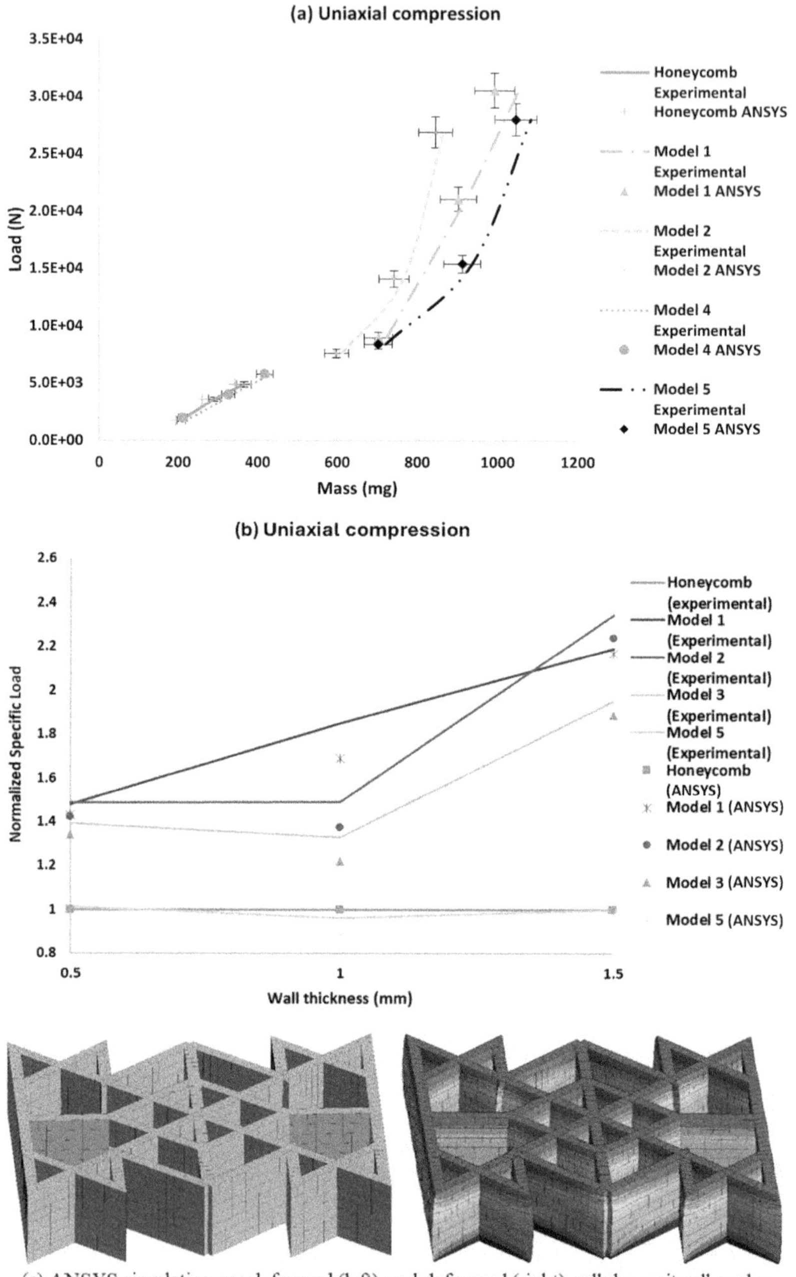

(c) ANSYS simulations undeformed (left) and deformed (right) cellular unit cell under uniaxial compression

FIGURE 6.11   (a) Experimental and simulation mass vs. load comparison for several cellular unit cells (models 1, 2, 4, and 5) under uniaxial compression, (b) experimental and simulation wall thickness vs. load for the several cellular unit cells (models 1, 2, 3, and 5), and (c) ANSYS simulation of a cellular unit cell.

6.6.1   Lattice Unit Cells

ANSYS simulations of a few sample fingerprints that satisfy the selection crite-rion and undergo uniaxial compression testing are presented in Figure 6.12. The results show that the predicted lattice unit cells exhibit different patterns of defor-mation under different ranges of compression loads (40–120%), while all of them exhibit superior performance as compared to the octet lattice. Although there is no single common factor that explains the superior performance of the predicted lattice unit cells over the octet lattice, some structural features such as low mass in lattices A and B; parallel orientation of truss members in lattices B, C, and D; and stronger joint connectivity in lattices C and D may account for their superior performance.

To improve upon the already optimized lattice unit cells, we will consider the fingerprints that outperform the octet unit cell. These superior structures will be used to create a new dataset of approximately 500 lattice unit cells that perform better than the octet unit cell. This latest dataset will then be fed into the discrimi-nator, and the generator will be trained to create new fingerprints by learning from this new dataset. The GAN system will generate novel fingerprints that are

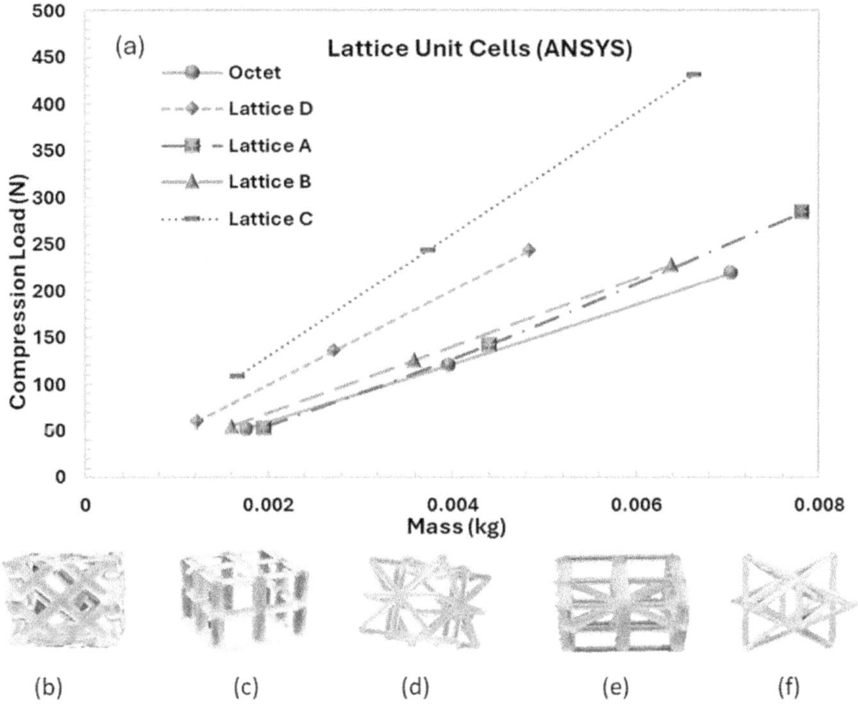

FIGURE 6.12   (a) ANSYS simulation comparisons for uniaxial compression of various optimal unit cells with respect to mass predicted through inverse design, (b) lattice A, (c) lattice B, (d) lattice C, (e) lattice D, and (f) octet lattice unit cell.

FIGURE 6.13 (A) several lattice cored sandwich structures, (b) lattice A, (c) lattice b, (d) lattice c, (e) lattice d, and (f) octet sandwich structures. The figure shows the 3D-printed unit cells using clear polymer.

closely aligned to this new sample space. These fingerprints can then be passed through the same set of boundary conditions.

The predicted lattice unit cells are used to prepare sandwich structures as the core. One potential application of lattice unit cells is to construct lattice-cored sandwich panels. In this chapter, the optimal lattice unit cells predicted by the inverse machine learning framework were used to construct several sandwich structures with varying densities. Lattice cored sandwich structures are constructed by sandwiching the lattice core in between two thin plates on the top and bottom. The core is made by stacking lattice unit cells side by side. Thin sheets with 10% of the thickness of the core are used to laminate the 4 by 4 unit cells, forming several sandwich structures with varying dimensions (see Figure 6.13 for several 3D printed lattice-cored sandwich panels). Testing procedures similar to the lattice unit cells were performed on the sandwich structures, and the results can be observed in the comparison of compression strength with densities in Figure 6.14. The compressive strengths of the sandwich structures are obtained by dividing the maximum compressive loads of the unit cells by the cross-sectional area. The sandwich structures performed in a similar pattern to those observed in the single-unit cell comparisons. Under uniaxial compression, the sandwich structures constructed using optimal lattice unit cells perform 60% better as compared to the octet lattice-cored sandwich structures.

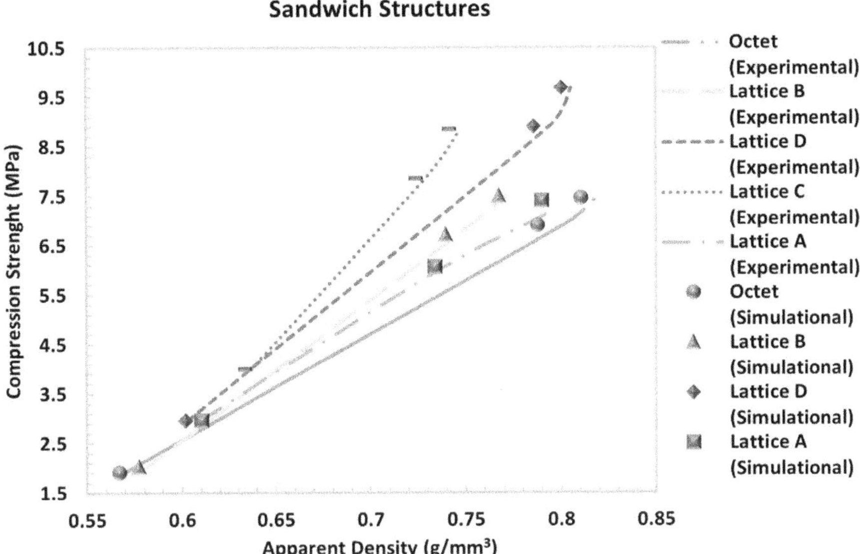

FIGURE 6.14 Compression strength vs. apparent density comparisons between ANSYS simulation and experimental testing of several lattice-cored sandwich panels in Figure 6.13.

In order to obtain further optimized structures, the target values for mass and compression strength will be set much higher than those from the previous learning cycles. The aim will be to obtain even lower mass and higher compressive strength than the octet unit cell. By doing so, further optimized lattice unit cells that perform even better than the octet unit cell can be predicted.

The inverse design framework in this study allows continuous optimization of the lattice unit cells by iteratively utilizing the framework in Figure 6.1. To quantify the capability of the inverse machine learning framework, several cycles of optimized lattice unit cells are generated by optimizing the structures for each cycle. Four sets or generations of these unit cells and their performances under uniaxial compression loading are presented in Figure 6.15. Initially, the boundary conditions of the inverse design framework are set to predict lattice unit cells that are better than the octet unit cell by constraining the mass and compressive strength of the predicted unit cells. As stated in the inverse machine learning section, each new set of optimal lattice unit cells obtained is generated by training the GANs with a new optimal dataset generated from the previous set. From the ANSYS simulation results in Figure 6.15, the improvement in the structural performance from the first generation (set 1) to the fourth generation (set 4) can be clearly observed to be increased by 50%. Figure 6.16 shows the images of the 16 optimal lattice unit cells. Table 6.1 summarizes the performance of the further optimized lattice unit cells.

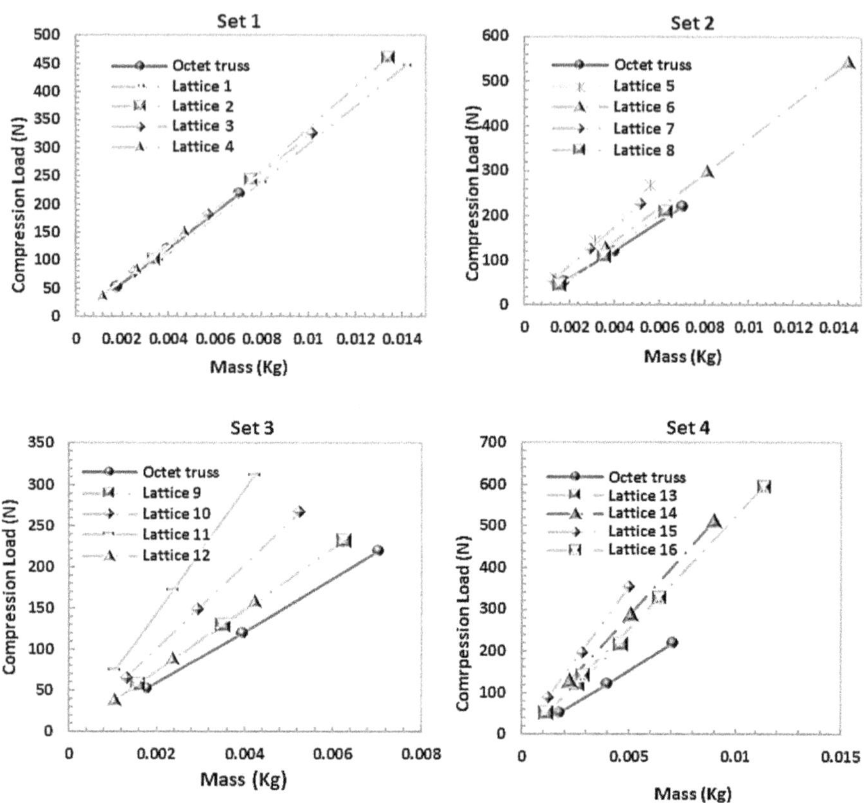

FIGURE 6.15 ANSYS simulations for compression load vs. mass of several further optimized lattice unit cells optimized for each generation (sets 1–4) (refer to Table 6.1 for more information on lattice unit cells).

From meticulous scrutiny of the mean alterations in mass and compression load percentages, as delineated in the tabulated data in Table 6.1, a discernible pattern emerges. It becomes evident that with each successive optimization iteration, denoted by distinct sets, there is a discernible enhancement in the mass, the compression load, or both. This observation underscores a progressive refinement in the performance metrics with each iteration of the optimization process.

The quantification of alterations in mass and compression load percentages for individual lattice unit cells is derived by juxtaposing these properties against the reference octet unit cell. The ensuing negative or positive signs associated with these differentials delineate the direction of change in comparison to the octet unit cell's properties. Specifically, a negative sign signifies a reduction in mass or compression load relative to the octet unit cell, whereas a positive sign denotes an augmentation in these properties. The advantageous implications of these alterations are multifaceted. A reduction in mass is particularly advantageous as it exercises control over the relative density, a pivotal parameter in the domain of lattice

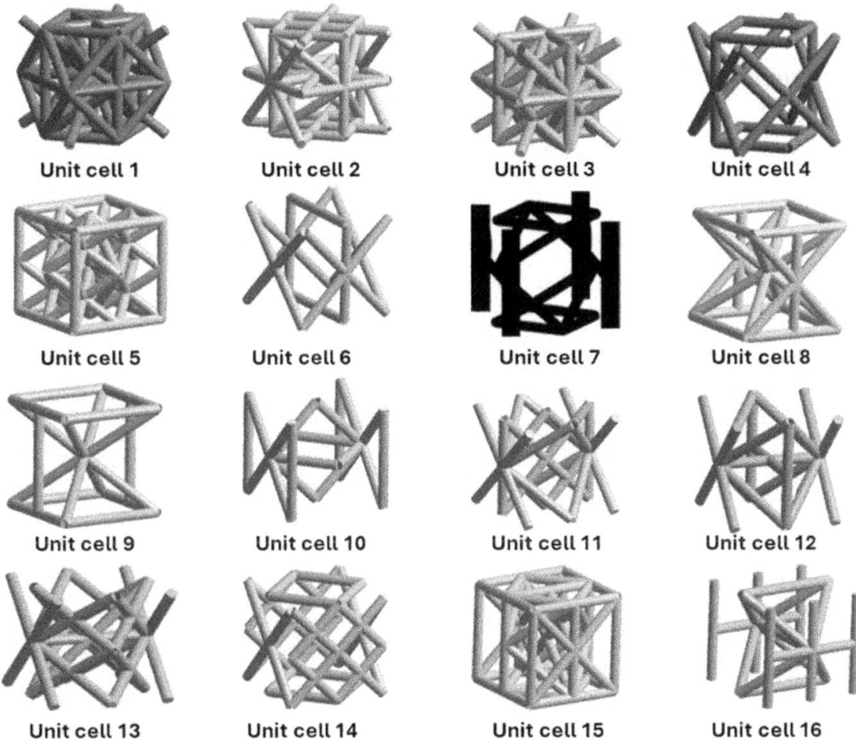

FIGURE 6.16   Sixteen further optimized lattice unit cells.

structures. Simultaneously, an augmentation in compression load bears a direct correlation to the compression strength of the unit cells. This intricate interplay between mass reduction and compression load increase encapsulates the essence of the optimization process, where the pursuit of enhanced mechanical performance is steered by a delicate balance between these fundamental structural attributes.

6.6.2 THIN-WALLED CELLULAR UNIT CELLS

6.6.2.1 Dynamic Mechanical Analysis

To observe the behavior of the predicted unit cells under cyclic loading, several designs were 3D printed using the same digital light processing (DLP) printer and clear photopolymer mentioned in the uniaxial compression section. An RSA-G2 dynamic mechanical analyzer is used to conduct dynamic analysis on all the specimens at a constant room temperature with varying frequency as shown in Figure 6.17. The storage modulus, loss modulus, and damping (tan(delta) or tanδ) properties of the specimen are obtained for an angular frequency range of 1–62 rad/sec (0.16–10Hz) and 0.125% strain under compression. All the specimens are designed to have similar overall volume (2 cm × 1.2 cm × 1.15 cm).

TABLE 6.1

Performance of the Further Optimized Lattice Unit Cells

Set 1	Mass (kg)	Change in Mass (%) Compared to Octet Lattice Cell	Compression Load (N)	Change in Compression Load (%) Compared to Octet Lattice Cell
Lattice 1	0.00117	−33.50196101	37.56716981	−27.81
	0.002632	−33.49838319	84.95471698	−0.29
	0.00468	−33.50053999	152.0150943	−0.31
Lattice 2	0.003534	100.8582959	100.5358491	93.20
	0.007951	100.8589329	238.5433962	99.04
	0.014135	100.861138	446	103.65
Lattice 3	0.003343	90.01307338	101.9622642	95.94
	0.007522	90.01364188	244.8075472	104.26
	0.013372	90.01875746	461.6226415	110.78
Lattice 4	0.002538	44.27897459	79.70943396	53.18
	0.005711	44.27799111	181.0867925	51.10
	0.010153	44.27613255	325.5811321	48.66
Set 1 summary	Average	32.18358101	Average	45.71

Set 2	Mass (kg)	Change in mass (%) compared to octet lattice cell	Compression load (N)	Change in compression load (%) compared to octet lattice cell
Lattice 5	0.0012844	−26.99369067	54.78113	5.27
	0.00289	−26.99070331	125.1245	4.40
	0.0051378	−26.99084863	226.3811	3.37
Lattice 6	0.001399	−20.47973626	60.1283	15.55
	0.0031476	−20.48302344	143.5019	19.74
	0.0055958	−20.4825783	268.5623	22.63
Lattice 7	0.0015638	−11.11237424	44.1434	−15.17
	0.0035186	−11.11054972	109.4075	−8.71
	0.0062553	−11.11095322	209.9396	−4.14
Lattice 8	0.0036154	105.5021884	128.3925	146.73
	0.0081346	105.5022231	297.3547	148.11
	0.014462	105.5078724	543.8113	148.31
Set 2 summary	Average	8.797364134	Average	30.38

Set 3	Mass (kg)	Change in mass (%) compared to octet lattice cell	Compression load (N)	Change in compression load (%) compared to octet lattice cell
Lattice 9	0.001561	−11.27721253	56.88301887	9.31
	0.003512	−11.27728375	128.9622642	7.60
	0.006244	−11.27721253	231.445283	5.68
Lattice 10	0.001058	−39.83970898	74.66037736	43.47
	0.002381	−39.8418553	171.1307953	42.79
	0.004233	−39.84255101	310.8	41.91

(Continued)

TABLE 6.1 *(Continued)*
Performance of the Further Optimized Lattice Unit Cells

Lattice 11	0.001311	−25.47604161	65.60754717	26.08
	0.00295	−25.47746564	148.6603774	24.04
	0.005244	−25.47746263	266.6075472	21.73
Lattice 12	0.001058	−39.83970898	38.6	−25.82
	0.002381	−39.8418553	87.88679245	−26.67
	0.004233	−39.84255101	158.554717	−27.60
Set 3 summary	Average	−21.83193183	Average	8.91

Set 4	Mass (kg)	Change in mass (%) compared to octet lattice cell	Compression load (N)	Change in compression load (%) compared to octet lattice cell
Lattice 13	0.0011462	−34.84908771	52.50943	0.91
	0.002579	−34.8474131	119.5057	−0.29
	0.0045849	−34.84766669	215.4566	−1.62
Lattice 14	0.0022451	27.61325527	127.3774	144.78
	0.0050515	27.61469281	287.2302	139.66
	0.0089804	27.61325527	512	133.78
Lattice 15	0.001239	−29.57426249	87.30189	67.77
	0.0027879	−29.57002829	197.4226	64.73
	0.0049562	−29.57142045	353.117	61.24
Lattice 16	0.0028358	61.1891093	141.6755	172.26
	0.0063806	61.19139046	327.3019	173.10
	0.011343	61.18626727	595.3962	171.86
Set 4 summary	Average	4.571755729	Average	70.53

Note: The table consists of different lattice unit cell names in column 1, mass in column 2, change in mass in column 3, compression load in column 4, and change in compression load compared to octet unit cell in column 5.

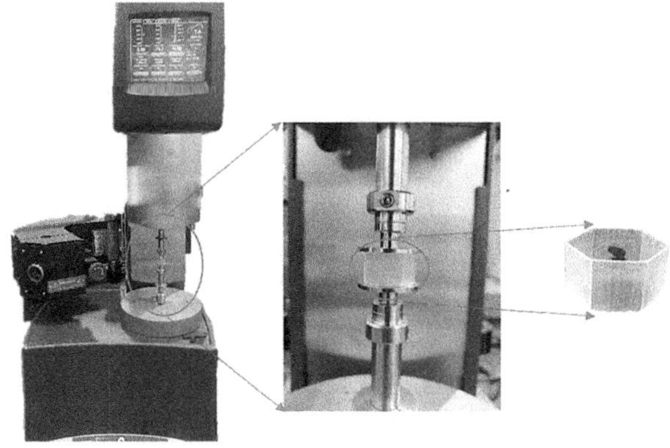

FIGURE 6.17   RSA-G2 dynamic mechanical analyzer with honeycomb unit cell specimen.

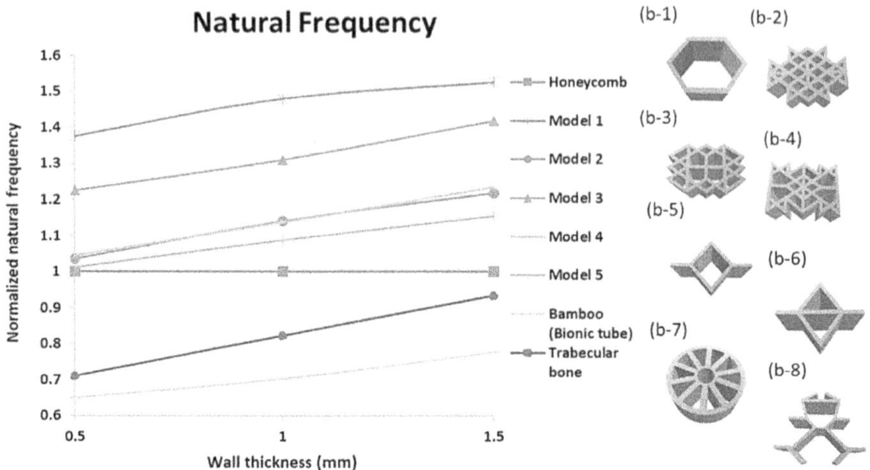

FIGURE 6.18 (a) Numerical simulations for wall thickness vs. normalized natural frequency of optimal cellular unit cells compared with biomimetic cellular unit cells and (b-1) honeycomb, (b-2) model 1, (b-3) model 2, (b-4) model 3, (b-5) model 4, (b-6) model 5, (b-7) bamboo (bionic tube), and (b-8) trabecular bone.

ANSYS workbench design modeler is used to design the cellular unit cells and the model analysis tool using mechanical APDL solver is employed to simulate the natural frequency of each cellular unit cell with fixed support at one end and uniform displacement of 20% at the other end for all the designs. The optimized unit cells are compared for their natural frequency. The comparisons are made by designing several unit cells with varying wall thickness and the calibrated normalized first natural frequencies. Here, honeycomb unit cells are considered as the datum structure; hence, the normalized natural frequencies are calculated with the ratio between the first natural frequency of the predicted unit cell over the first frequency of the honeycomb unit cell.

It can be seen from Figure 6.18 that the normalized natural frequency of the optimized cellular unit cells is about 10–50% higher than the biomimetic unit cells like honeycomb, bamboo, and trabecular bone structures that are widely studied for their high natural frequency and energy absorption properties. The honeycomb structure can be seen to perform better than other biomimetic structures like the bamboo stem and trabecular bone within the same overall volume. It is seen that within the same mass range, models or unit cells 1–4 all have a higher natural frequency than that of the biomimetic counterparts. Once the mass exceeds about 2g, model 5 also shows higher natural frequency. Figure 6.19 shows the mass vs. natural frequency comparisons. The higher natural frequency makes models 1–5 a better choice as cellular unit cells. Hence, these structures are considered for further experimental and simulation validations to observe their behavior under uniaxial compression, DMA, and impact tests.

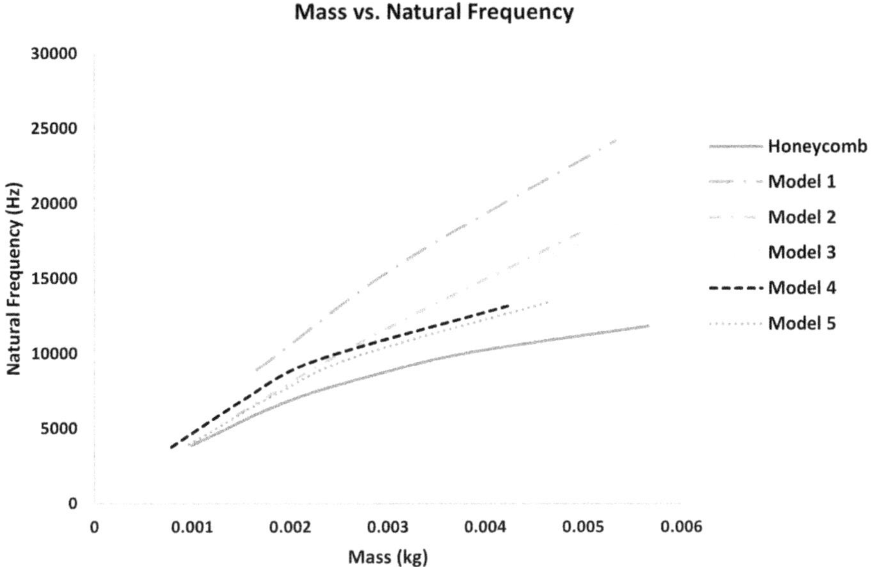

FIGURE 6.19   Mass vs. natural frequency comparison of the discovered cellular unit cells.

Figure 6.20(a) shows the storage modulus of different optimized cellular structures compared to honeycomb structures with frequency sweep at room temperature. It can be observed that the storage modulus of the cellular structures trend to gradually increase with higher frequency. Figure 6.20(b) shows the variations in normalized tanδ with wall thicknesses for different optimal cellular unit cell designs compared to honeycomb unit cells. The normalized tan δ in Figure 6.20(b) is obtained in a similar manner to the normalized specific load in the uniaxial compression comparisons, i.e., the normalized tanδ is a ratio of the (tanδ/specific density) of each discovered unit cell over the (tanδ/specific density) of the honeycomb unit cell.

The optimal cellular unit cells like models 1, 2, and 3 can be seen to perform decently in comparison to honeycomb unit cells in terms of tanδ. It can be observed that all the structures follow similar trends and the storage modulus and tanδ increase with mass and frequency, except for model 1 in storage modulus and model 3 in tanδ. With the same overall volume, structures like models 1 and 2 exhibit a wider range of damping properties because of their higher natural frequencies and mass. Figure 6.21(a) and (b) show the storage modulus and tanδ comparisons.

The deformation modes of a unit cell using ANSYS simulation are shown in Figure 6.22.

6.6.2.2   Low Velocity Impact Analysis

The optimal cellular unit cells derived from the inverse design framework have also been employed in fabricating sandwich structures with cellular cores. These

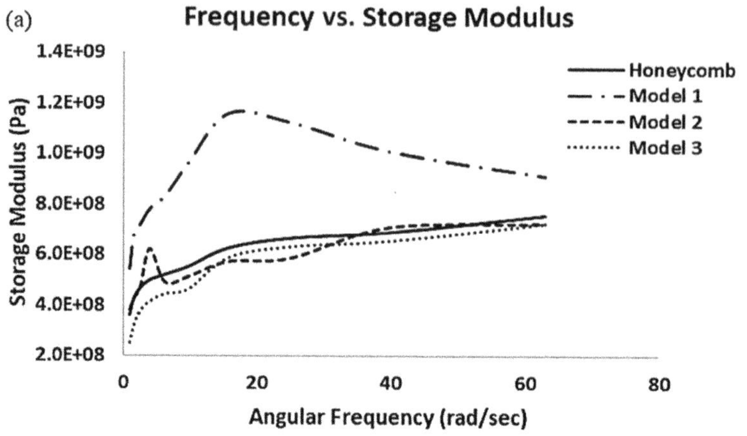

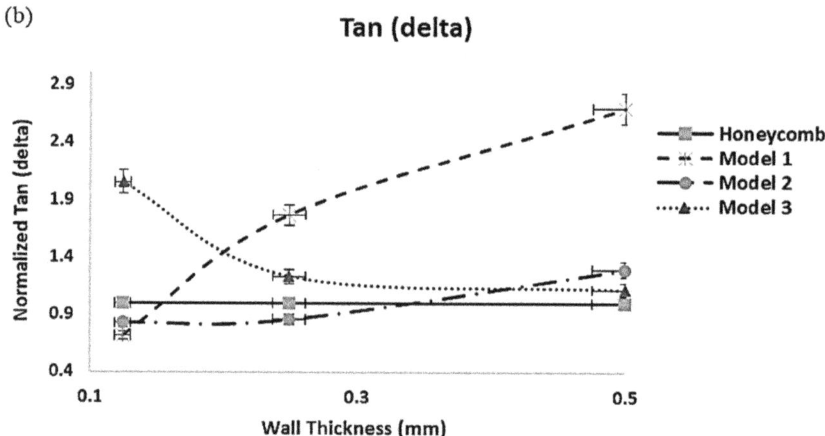

FIGURE 6.20 (a) Storage modulus plotted with angular frequency and (b) normalized tanδ plotted with wall thickness for different cellular structures.

structures are manufactured using extrusion-based 3D printing technology, utilizing polylactic acid (PLA) as the ink. The manufacturing process is demonstrated in Figure 6.23(b). To assess the impact resistance of the sandwich structures, low velocity impact tests were conducted utilizing an Instron Dynatup 8250 H V impact tester, with a hammer weight of 11.2 kg and impact velocity of 2 m/s, as depicted in Figure 6.23(a). The sandwich structures, which have a consistent overall volume of 120 mm × 25.4 mm × 4 mm, were designed using SolidWorks software. Explicit nonlinear finite element simulations of the low velocity impact tests were conducted utilizing ANSYS LS-DYNA software. These simulations enable the prediction of the structural response of the sandwich structures to the impact, providing insight into their deformation and damage mechanisms. The

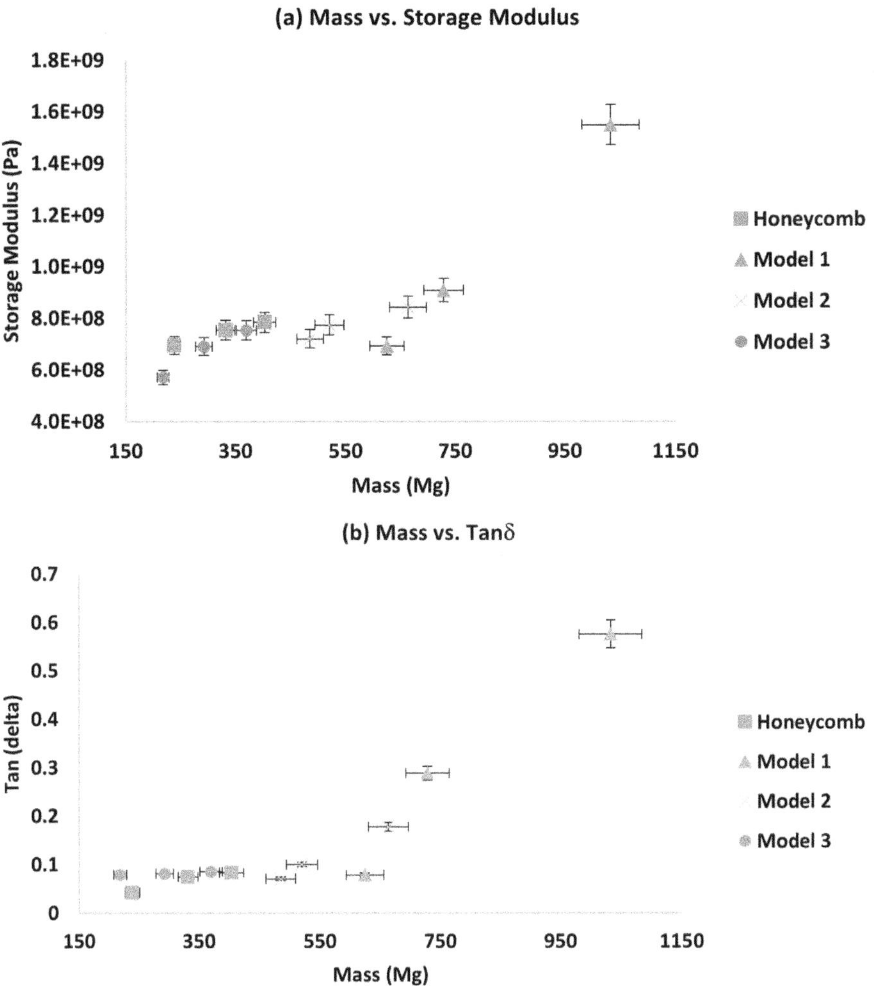

FIGURE 6.21 (a) Mass vs. modulus and (b) mass vs. tanδ comparisons of discovered cellular unit cells (models 1, 2, and 3) with the control honeycomb unit cell.

results obtained from the simulations can be used to optimize the design of the sandwich structures to enhance their impact resistance properties.

Figure 6.24(a) illustrates that when the cellular structures are positioned flat, with their walls perpendicular to the surface, they exhibit a significant improvement of 300–800% in terms of normalized energy compared to the honeycomb structure. This enhanced performance is attributed to the absence of substantial voids or porosity in the optimal cellular unit cells, unlike the honeycomb structures. Instead, the optimal cellular unit cells feature intricate web-like designs that effectively cover the impact region of the sandwich structure. In vertical orientations, where the walls of the structures are parallel to the surface, the optimal

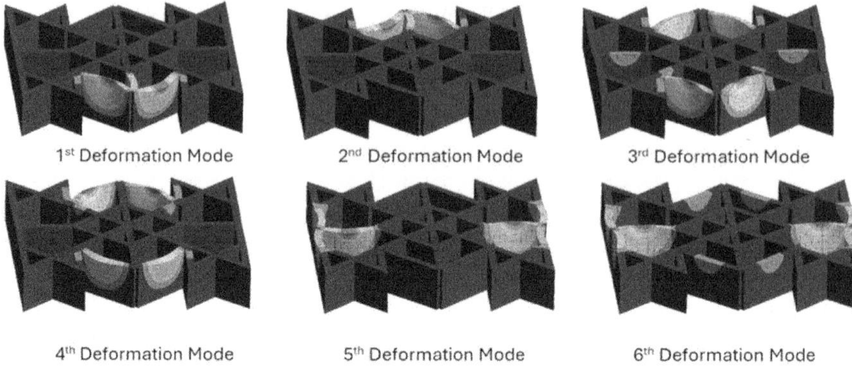

FIGURE 6.22   Deformation mode of a unit cell using NASYS simulation.

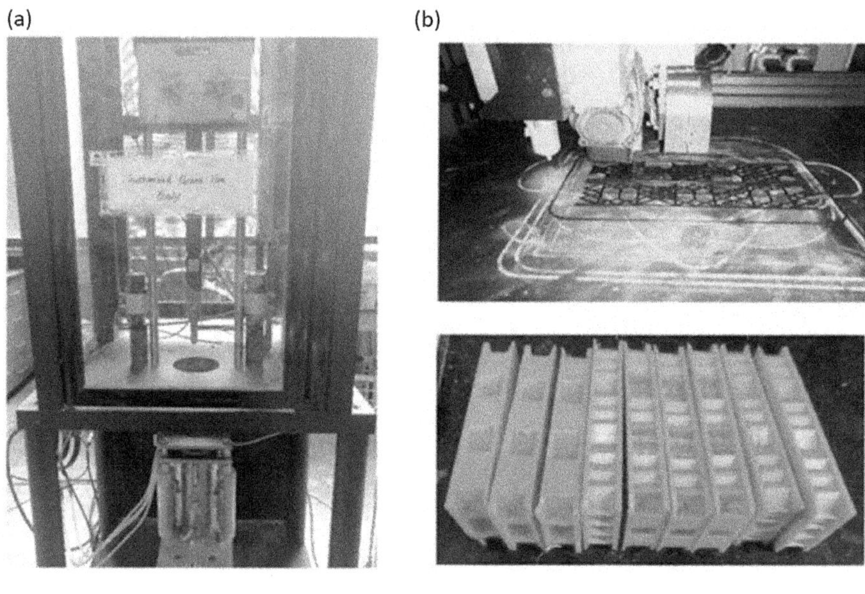

FIGURE 6.23   (a) Instron Dynatup impact tester and (b) extrusion-based 3D printer (upper right) and additively manufactured sandwich structures (lower right).

structures demonstrate a performance level ranging from 50 to 350% of the honeycomb structure (Figure 6.24(b)). Notably, experimental tests (Figure 6.24(c)) conducted on sandwich structures constructed with the optimal cellular lattice core provide evidence of the superior performance of these structures. Figure 6.24(c) also reveals a gradual decrease in the normalized energies of the optimal sandwich panels as the wall thickness of the unit cells increases. This decline occurs because an increase in the wall thickness of the unit cells leads to a corresponding increase in the mass of the structures. However, since the impact is localized at the central unit cell of the sandwich structure, the additional mass in the

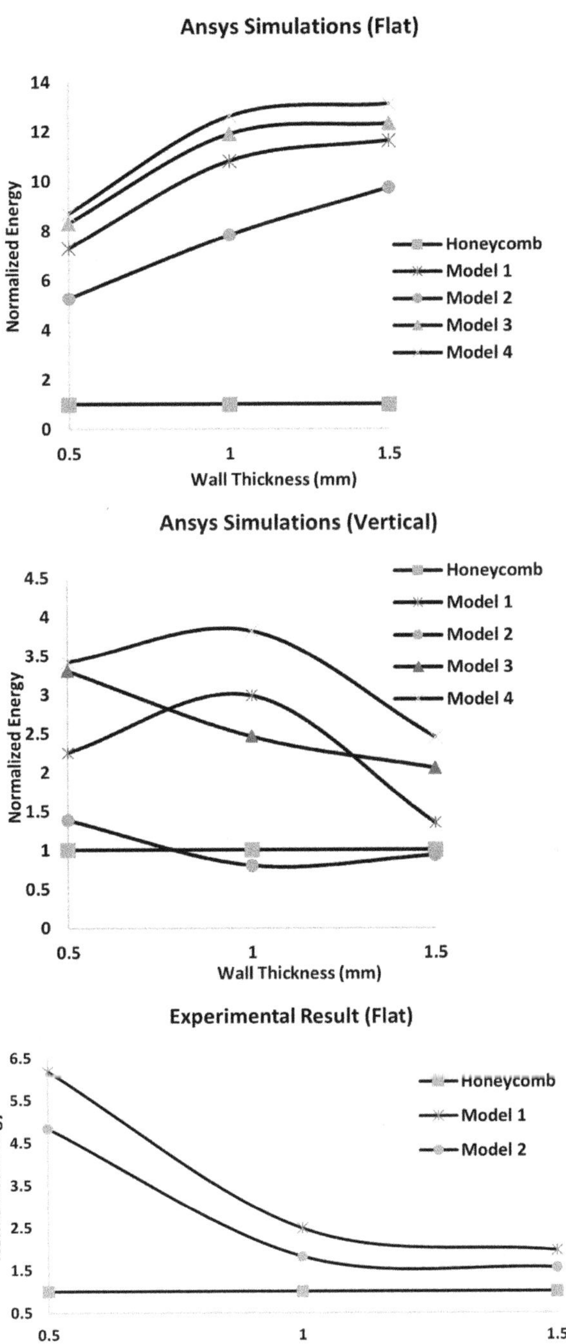

FIGURE 6.24 Numerical simulations for low velocity impact test on various optimal cellular structures in (a) flat, (b) vertical orientations, and (c) experimental results in a flat orientation.

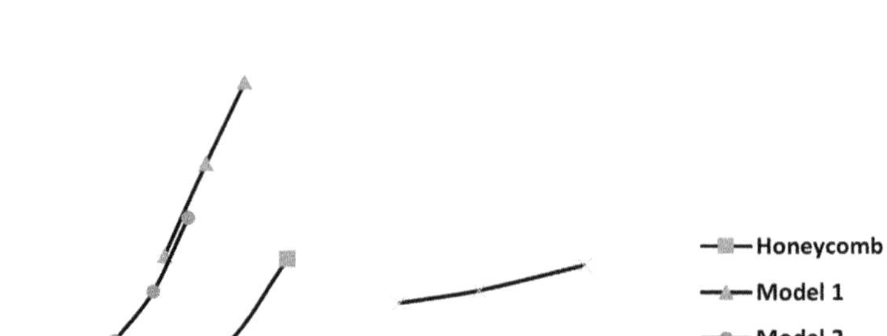

FIGURE 6.25 Mass vs. energy, simulation result curves comparing honeycomb, solid, model 1, and model 2 unit cells. The curves for models 1 and 2 show superior energy values.

remaining unit cells within the sandwich structure does not contribute to the overall performance (normalized energy) of the sandwich structure. Consequently, the augmented mass of the sandwich structure negatively affects the normalized energy when compared to wall thickness.

It is important to note that the experimental results for the impact test in the vertical orientation are not presented in this study, as the optimization through machine learning only focuses on unit cells in the flat orientation. The difference observed in the normalized energies between the numerical and experimental analyses in the flat direction (Figure 6.24(a) and (c)) is attributable to the varying number of unit cells used in the sandwich structures in each method. The analytical study employs a single cellular unit cell to balance computational time and power during dynamic analysis, whereas the experimental study employs multiple unit cells to meet the minimum dimensional requirements for the specimens (120 mm × 25.4 mm × 4 mm) when utilizing the impact testing machine. More comparisons can be seen in Figure 6.25.

The deformation mode of a unit cell during low velocity impact using ANSYS simulation is shown in Figure 6.26.

6.6.2.3 Effect of Porosity on the Mechanical Strength of the Cellular Unit Cells

In essence, cellular unit cells belong to porous materials. It is well-known that porosity plays a critical role in the load-carrying capacity of the porous materials. Figure 6.27 delves into porosity vs. normalized load comparisons, revealing a compelling trend. Across the porosity range of 90–98%, optimal unit cells

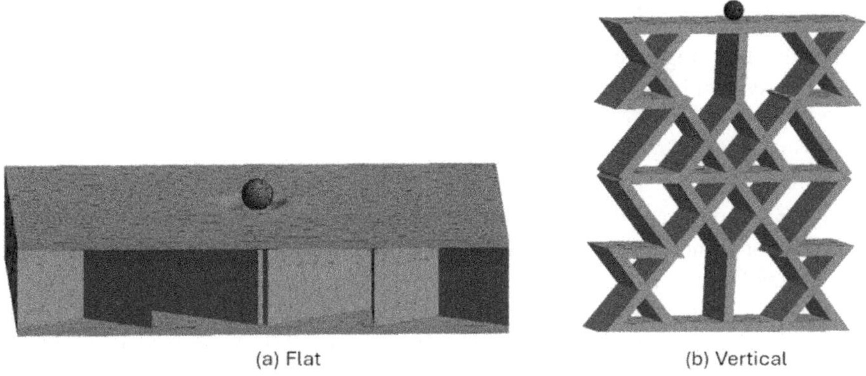

FIGURE 6.26   Deformation mode of a unit cell under low velocity impact using ANSYS simulation. (a) Flat and (b) vertical impacts.

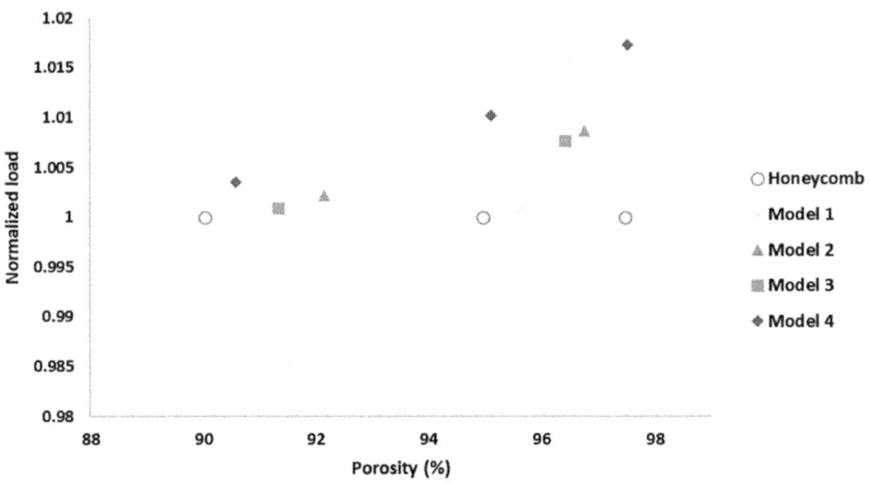

FIGURE 6.27   Porosity vs. normalized load.

consistently highlight superior load-carrying capacities compared to the honeycomb unit cell, with the exception of model 1 at a porosity of 90%.

6.7 THEORETICAL BACKGROUND

Now, we provide some theoretical background for the natural frequency and effective modulus calculation.

6.7.1 Natural Frequency

From the Lagrange's equations, the equation of motion for a multi-degree-of-freedom system in matrix form can be derived as follows:

$$[m]\ddot{\vec{x}}+[k]\vec{x} = \vec{F}, \tag{6.4}$$

where $[m]$ and $[k]$ are the mass and stiffness matrices, respectively; $\vec{F}$ is the column vector of nonconservative generalized force; and $\vec{x}$ is the column vector of generalized velocity.

Now, the solution of the equation of motion for a conservative system corresponds to the undamped free vibration of a system [75]. Assuming a solution of the form $x_i(t) = X_i T(t)$, $i = 1, 2, \ldots, n$, where X_i is a constant and T is a function of time t, and substituting it in Equation 6.4 gives:

$$[m]\vec{X}\ddot{T}(t)+[k]\vec{X}T(t) = 0. \tag{6.5}$$

Equation 6.5 can be written in scaler form with n separate equations as follows:

$$\left(\sum_{j=1}^{n}m_{ij}X_j\right)\ddot{T}(t)+\left(\sum_{j=1}^{n}k_{ij}X_j\right)T(t) = 0, \; i = 1, 2, \ldots, n. \tag{6.6}$$

Equation 6.6 will give the following relations:

$$-\frac{\ddot{T}(t)}{T(t)} = \frac{\left(\displaystyle\sum_{j=1}^{n}k_{ij}X_j\right)}{\left(\displaystyle\sum_{j=1}^{n}m_{ij}X_j\right)}, \; i = 1, 2, \ldots, n. \tag{6.7}$$

Since Equation 6.7 is independent of the index i and t, both sides must be equal to a constant ω^2, which leads to the following:

$$\ddot{T}(t)+\omega^2 T(t) = 0, \tag{6.8}$$

$$\sum_{j=1}^{n}(k_{ij}-\omega^2 m_{ij})X_j = 0, \; i = 1, 2, \ldots, n, \tag{6.9}$$

$$\left[[k]-\omega^2[m]\right]\vec{X} = \vec{0}. \tag{6.10}$$

To obtain a nontrivial solution, the determinant of the coefficient matrix of Equation 6.10 must be zero. This leads to the following:

$$\Delta = \left|k_{ij}-\omega^2 m_{ij}\right| = \left\|[k]-\omega^2[m]\right\| = 0, \tag{6.11}$$

where ω^2 is the eigenvalue or the characteristic value and ω is the natural frequency of the system.

6.7.2 GENERALIZED HOOKE'S LAW FOR EFFECTIVE PROPERTIES OF CELLULAR STRUCTURES

The general case of stress–strain components generalized by Hooke's law contains 81 elastic constants which are reduced to 36 due to the symmetry of the stress and strain tensors. This gives the stress–strain relation of the anisotropic body as follows [76–79]:

$$
\begin{pmatrix} \bar{\sigma}_{11} \\ \bar{\sigma}_{22} \\ \sigma_{33} \\ \bar{\sigma}_{23} \\ \bar{\sigma}_{13} \\ \bar{\sigma}_{12} \end{pmatrix} = \begin{pmatrix} C_{11} & C_{12} & C_{13} & C_{14} & C_{15} & C_{16} \\ C_{21} & C_{22} & C_{23} & C_{24} & C_{25} & C_{26} \\ C_{31} & C_{32} & C_{33} & C_{34} & C_{35} & C_{36} \\ C_{14} & C_{24} & C_{34} & C_{44} & C_{45} & C_{46} \\ C_{15} & C_{25} & C_{35} & C_{45} & C_{55} & C_{56} \\ C_{16} & C_{26} & C_{36} & C_{46} & C_{56} & C_{66} \end{pmatrix} \begin{pmatrix} \bar{\varepsilon}_{11} \\ \bar{\varepsilon}_{22} \\ \bar{\varepsilon}_{33} \\ \bar{\gamma}_{23} \\ \bar{\gamma}_{13} \\ \bar{\gamma}_{12} \end{pmatrix},
\tag{6.12}
$$

where the $6 \times 6[C]$ matrix is the stiffness matrix.

For orthotropic structures where three mutually perpendicular planes are in symmetry, the stiffness matrix can be reduced to 9 components as $(C_{14}, C_{15}, C_{16}, C_{24}, C_{25}, C_{26}, C_{34}, C_{35}, C_{36}, C_{45}, C_{46}, C_{56}) = 0$ due to symmetry. The stiffness matrix of an orthogonal structure is of the following form:

$$
\begin{pmatrix} \bar{\sigma}_{11} \\ \bar{\sigma}_{22} \\ \sigma_{33} \\ \bar{\sigma}_{23} \\ \bar{\sigma}_{13} \\ \bar{\sigma}_{12} \end{pmatrix} = \begin{pmatrix} C_{11} & C_{12} & C_{13} & 0 & 0 & 0 \\ C_{21} & C_{22} & C_{23} & 0 & 0 & 0 \\ C_{31} & C_{32} & C_{33} & 0 & 0 & 0 \\ 0 & 0 & 0 & C_{44} & 0 & 0 \\ 0 & 0 & 0 & 0 & C_{55} & 0 \\ 0 & 0 & 0 & 0 & 0 & C_{66} \end{pmatrix} \begin{pmatrix} \bar{\varepsilon}_{11} \\ \bar{\varepsilon}_{22} \\ \bar{\varepsilon}_{33} \\ \bar{\gamma}_{23} \\ \bar{\gamma}_{13} \\ \bar{\gamma}_{12} \end{pmatrix}.
\tag{6.13}
$$

The strain tensor components are as follows:

$$
\bar{\varepsilon}_{11} = \frac{\partial u}{\partial x}, \quad \bar{\varepsilon}_{22} = \frac{\partial v}{\partial y}, \quad \bar{\varepsilon}_{33} = \frac{\partial w}{\partial z}, \quad 2\bar{\varepsilon}_{23} = \frac{\partial w}{\partial y} + \frac{\partial v}{\partial z},
$$

$$
2\bar{\varepsilon}_{13} = \frac{\partial w}{\partial x} + \frac{\partial u}{\partial z}, \quad 2\bar{\varepsilon}_{23} = \frac{\partial v}{\partial x} + \frac{\partial u}{\partial y},
\tag{6.14}
$$

where u, v, and $w =$ displacements in the x, y, and z directions.

6.8 SUMMARY

This chapter delves into the application of inverse machine learning techniques in predicting structural properties and designing lattice unit cells and cellular unit cells with enhanced mechanical properties. In previous chapters, forward machine learning techniques were employed, but their limitations prompted the exploration of an inverse framework utilizing GANs. This novel framework aims to predict optimal structures given desired properties, marking a significant advancement in material science and engineering.

Because the forward design model was presented in Chapter 5 for lattice unit cells, this chapter focuses on the forward design model for cellular unit cells and especially the inverse machine learning frameworks for the discovery and optimization of lattice unit cells and cellular unit cells.

In the forward machine learning approach, predictive models were developed to forecast the mechanical properties of various cellular unit cells, focusing on thin-walled structures with superior natural frequencies. Through rigorous training and evaluation, regression models such as GPR and SVMs were found to accurately predict mass, load, and natural frequency. These models provided a foundation for subsequent analysis and optimization.

The inverse design framework using GANs represents a paradigm shift in structural optimization. By inputting desired properties, such as low mass and high compression strength, the framework generates optimal lattice unit cells. This process involves training a discriminator to differentiate between real and generated data, with iterative refinement of the generator to produce increasingly accurate results. The resulting lattice unit cells exhibit superior performance compared to traditional designs such as octet unit cells, demonstrating the efficacy of the inverse design approach.

Validation of the optimized lattice unit cells and cellular unit cells was conducted through numerical simulations and experimental tests under uniaxial compression, dynamic mechanical analysis, and low velocity impact. The results showed a significant improvement in structural performance, with the optimized designs outperforming traditional lattice structures such as octet unit cells and thin-walled structures such as honeycomb unit cells. Furthermore, the frameworks facilitated the design of high-performance sandwich structures, leveraging the optimized lattice unit cells and thin-walled cellular unit cells to enhance overall strength and durability. The inverse design framework using GANs holds immense promise for revolutionizing structural optimization in material science and engineering. By leveraging machine learning techniques, researchers can expedite the design process, optimize performance, and unlock new possibilities in the development of advanced materials and structures. This chapter also provides a comprehensive overview of this innovative approach and its potential implications for future research and industrial applications.

REFERENCES

1. Gao H, Ji T, Lin W, Guo X, Zhang T. Forward machine learning for designing biomimetic rods with superior buckling resistance. Journal of Mechanical Design, 142:081701, (2020).
2. Lin W, Ji T, Gao H, Guo X, Zhang T. Optimizing lattice structures by forward machine learning. Journal of Mechanical Design, 142:091703, (2020).
3. Chen K, Chen Y, Li K, Ma J, Wang L. Deep learning for materials design: A review. Journal of Materials Science, 56:2161–2190, (2021).
4. Jin L, Zhou Y, Chen J, Huang L. Recent advances in inverse design of materials using deep generative models. Computational Materials Science, 162:121–131, (2019).
5. Teimouri A, Alinia M, Kamarian S, Saber-Samandari S, Li G, Song JI. Design optimization of additively manufactured sandwich beams through experimentation, machine learning, and imperialist competitive algorithm. Journal of Engineering Design, 35:320–337, (2024).
6. Challapalli A, Li G. 3D printable biomimetic rod with superior buckling resistance designed by machine learning. Scientific Reports, 10:20716, (2020).
7. Challapalli A, Li G. Machine learning assisted design of new lattice core for Sandwich structures with superior load carrying capacity. Scientific Reports, 11:18552, (2021).
8. Gómez-Bombarelli R, Wei JN, Duvenaud D, Hernández-Lobato JM, Sánchez-Lengeling B, Sheberla D, Aguilera-Iparraguirre J, Hirzel TD, Adams RP, Aspuru-Guzik A. Automatic chemical design using a data-driven continuous representation of molecules. ACS Central Science, 4:268–276, (2018).
9. Ma Q, Li Y, Li Z, Li B, Liu Y. Inverse design of nanostructures with machine learning: A review. Nanoscale, 12:18137–18153, (2020).
10. Kang Y, Lee J, Kim H, Kim J. Generative adversarial networks for material science. Computational Materials Science, 158:460–471, (2019).
11. Wang H, Li J, Zhang Y, Li Y, Li L. Design of metamaterials by inverse machine learning: A review. Materials, 14:2668, (2021).
12. Ma J, Yang Z, Zhou L, Wang J. Inverse design of nanostructured materials using generative models. Physical Chemistry Chemical Physics, 22:24720–24730, (2020).
13. Guo X, Huang B, Zhang T. Design of lattice materials with optimal mechanical properties. Extreme Mechanics Letters, 14:30–40, (2017).
14. Ma Q, Yan C, Chen C, Huang B, Zhang T. Inverse design of nanophotonic structures using complementary convex optimization. Journal of Optics, 21:035102, (2019).
15. Molesky S, Lin Z, Piggott AY, Jin W, Vucković J, Rodriguez AW. Inverse design for nanophotonics. Nature Photonics, 12:659–670, (2018, (2018).
16. Teimouri A, Challapalli A, Konlan J, Li G. Machine learning assisted design and optimization of plate-lattice structures with superior specific recovery force. Giant, 18:100282, (2024).
17. Goodfellow I, Pouget-Abadie J, Mirza M, Xu B, Warde-Farley D, Ozair S, Courville A, Bengio Y. Generative adversarial networks. Advances in Neural Information Processing Systems 63: 139–144, (2014).
18. Shorten C, Khoshgoftaar TM. A survey on image data augmentation for deep learning. Journal of Big Data, 6:60, (2019).
19. Minaee S, Boykov YY, Porikli F, Plaza AJ, Kehtarnavaz N, Terzopoulos D. Image segmentation using deep learning: A survey. IEEE Transactions on Pattern Analysis and Machine Intelligent, 44:3523–3542, (2023).

20. Karras T, Laine S, Aila T. A style-based generator architecture for generative adversarial networks. IEEE Transactions on Pattern Analysis and Machine Intelligent, 43:4217–4228, (2021).

21. Pouyanfar S, Sadiq S, Yan YL, Tian HM, Tao YD, Reyes MP, Shyu ML, Chen SC, Iyengar SS. A survey on deep learning: Algorithms, techniques, and applications. ACM Computing Surveys, 51:92, (2019).

22. Borji A. Pros and cons of GAN evaluation measures. Computer Vision and Image Understanding, 179:41–65, (2019).

23. Pan ZQ, Yu WJ, Yi XK, Khan A, Yuan F, Zheng YH. Recent progress on generative adversarial networks (GANs): A survey. IEEE Access, 7:36322–36333, (2019).

24. Shao SY, Wang P, Yan RQ. Generative adversarial networks for data augmentation in machine fault diagnosis. Computers in Industry, 106:85–93, (2019).

25. Lloyd S, Weedbrook C. Quantum generative adversarial learning. Physical Review Letters, 121:040502, (2018).

26. Wang ZR, Wang J, Wang YR. An intelligent diagnosis scheme based on generative adversarial learning deep neural networks and its application to planetary gearbox fault pattern recognition. Neurocomputing, 310:213–222, (2018).

27. Zhang W, Li X, Jia XD, Ma H, Luo Z, Li X. Machinery fault diagnosis with imbalanced data using deep generative adversarial networks. Measurement, 152:107377, (2020).

28. Fiore U, De Santis A, Perla F, Zanetti P, Palmieri F. Using generative adversarial networks for improving classification effectiveness in credit card fraud detection. Information Science, 479:448–455, (2019).

29. Shen YJ, Yang CY, Tang XO, Zhou BL. InterFaceGAN: Interpreting the disentangled face representation learned by GANs. IEEE Transactions on Pattern Analysis and Machine Intelligent, 44:2004–2018, (2022).

30. Gui J, Sun ZA, Wen YG, Tao DC, Ye JP. A review on generative adversarial networks: Algorithms, theory, and applications. IEEE Transactions on Knowledge and Data Engineering, 35:3313–3332, (2023).

31. Hong Y, Hwang U, Yoo J, Yoon S. How generative adversarial networks and their variants work: An overview. ACM Computing Survey, 52:10, (2019).

32. Wu F. English Vocabulary learning aid system using digital twin wasserstein generative adversarial network optimized with jelly fish optimization algorithm. Applied Artificial Intelligence, 38:2327908, (2024).

33. Li F, Zhang JL, Li KW, Peng Y, Zhang HT, Xu YP, Yu Y, Zhang YT, Liu ZW, Wang Y, Huang L, Zhou FF. GANSamples-ac4C: Enhancing ac4C site prediction via generative adversarial networks and transfer learning. Analytical Biochemistry, 689:115495, (2024).

34. Cheng XW, Huang K, Zou Y, Ma SJ. SleepEGAN: A GAN-enhanced ensemble deep learning model for imbalanced classification of sleep stages. Biomedical Signal Processing and Control, 92:106020, (2024).

35. Kong Q, Shibuta Y. Predicting materials properties with generative models: Applying generative adversarial networks for heat flux generation. Journal of Physics-Condensed Matter, 36:195901, (2024).

36. Yu ZR, Cui W. Robust hyperspectral image classification using generative adversarial networks. Information Science, 666:120452, (2024).

37. Zhang FL, Wang ZP, Wang QF, Ji QC. Predicting thermal stress in binary composites through advanced generative adversarial networks. MRS Communications, 14:397–401, (2024).

38. Huynh N, Deshpande G. A review of the applications of generative adversarial networks to structural and functional MRI based diagnostic classification of brain disorders. Frontiers in Neuroscience, 18:1333712, (2024).

39. Wang X, Mi Y, Zhang X. 3D human pose data augmentation using generative adversarial networks for robotic-assisted movement quality assessment. Frontiers in Neurorobotics, 18:1371385, (2024).
40. Dannehl M, Delouille V, Barra V. An experimental study on EUV-to-magnetogram image translation using conditional generative adversarial networks. Earth and Space Science, 11:e2023EA002974, (2024).
41. Zhang CH, Yu S, Tian ZY, Yu JJQ. Generative adversarial networks: A survey on attack and defense perspective. ACM Computing Surveys, 56:91, (2024).
42. Alshammari K, Alshammari R, Alshammari A, Alkhudaydi T. An improved pear disease classification approach using cycle generative adversarial network. Scientific Reports, 14:6680, (2024).
43. Zhao PH, Ding ZJ, Li Y, Zhang XH, Zhao YQ, Wang HJ, Yang Y. SGAD-GAN: Simultaneous generation and anomaly detection for time-series sensor data with generative adversarial networks. Mechanical Systems and Signal Processing, 210:111141, (2024).
44. Yan C, Feng X, Li G. From drug molecules to thermoset shape memory polymer: A machine learning approach. ACS Applied Materials & Interfaces, 13:60508–60521, (2021).
45. Karras T, Laine S, Aila TA Style-Based Generator Architecture for Generative Adversarial Networks. arXiv preprint arXiv:1812.04948, (2018).
46. Nie D, Zhang H, Adeli E, Liu L, Shen D. 3D deep learning for multi-modal imaging-guided survival time prediction of brain tumor patients. Medical Image Analysis, 43:128–139, (2018).
47. Zhao J, Song Y, Li Y, Li X, Li H. Adversarial learning-based anomaly detection on hyperspectral images. IEEE Transactions on Geoscience and Remote Sensing, 57:6548–6561, (2019).
48. Akcay S, Atapour-Abarghouei A, Breckon TP. Ganomaly: Semi-Supervised Anomaly Detection via Adversarial Training. Proceedings of the IEEE Conference on Computer Vision and Pattern Recognition Workshops, 68–77, (2019).
49. Zhu JY, Park T, Isola P, Efros AA. Unpaired Image-to-Image Translation Using Cycle-Consistent Adversarial Networks. Proceedings of the IEEE International Conference on Computer Vision, 2242–2251, (2017).
50. Korshunova I, Myslinskii M, Guha T, Boukhelifa N, Seidel HP. Fast and deep deformation approximations. ACM Transactions on Graphics, 37:137, (2018).
51. Arik SO, Chrzanowski M, Coates A, Firat A, Hestness J, Sengupta S. GANs trained by a two time-scale update rule converge to a local nash equilibrium. Advances in Neural Information Processing Systems, 31:700–710, (2019).
52. Challapalli A, Patel D, Li G. Inverse machine learning framework for optimizing lightweight metamaterials. Materials and Design, 208:109937, (2021)
53. Challapalli A, Konlan J, Patel D, Li G. Discovery of cellular unit cells with high natural frequency and energy absorption capabilities by an inverse machine learning framework. Frontiers in Mechanical Engineering, 7:779098, (2021).
54. Phi NH, Bui HN, Moon SY, Lee JW. Controlled synthesis of space-time modulated metamaterial for enhanced nonreciprocity by machine learning. Advanced Intelligent Systems, 6:2300565, (2024).
55. Liu S, Acar P. Generative adversarial networks for inverse design of two-dimensional spinodoid metamaterials. AIAA Journal, 62:2433–2442, (2024).
56. Tran T, Amirkulova FA, Khatami E. Broadband acoustic metamaterial design via machine learning. Journal of Theorical and Computational Acoustics, 30:2240005, (2022).
57. Qian C, Tan RK, Ye WJ. Design of architectured composite materials with an efficient, adaptive artificial neural network-based generative design method. Acta Materialia, 225:117548, (2022).

58. Guo BY, Deng L, Zhang HT. Non-local generative machine learning-based inverse design for scattering properties. Optics Express, 31:20872–20886, (2023).

59. Zhang D, Qin A, Shen S, Trinchi A, Lu GX. Energy absorption analysis of origami structures based on small number of samples using conditional GAN. Thin-Walled Structures, 188:110772, (2023).

60. Maxwell JC. On the calculation of the equilibrium and stiffness of frames. Philosophical Magazine, 27:294, (1864).

61. Evans AG, Hutchinson JW, Ashby MF. Multifunctionality of cellular metal systems. Progress in Materials Science, 43:171–221, (1998).

62. Deshpande VS, Fleck NA, Ashby MF. Foam topology bending vs stretching dominated architecture. Acta Materialia, 49:1035–1040, (2001).

63. Deshpande VS, Fleck NA, Ashby MF. Effective properties of the octet-truss lattice material. Journal of the Mechanics and Physics of Solids, 49:1747–1769, (2001).

64. Gibson LJ, Ashby MF. Cellular Solids: Structure and Properties, (2nd ed.). Cambridge University Press, Cambridge, MA, (1997).

65. Gibson JL, Ashby FM. The mechanics of three-dimensional cellular materials. Proceedings of the Royal Society of London A, 382:43–59, (1982).

66. Gibson JL. Cellular solids. MRS Bulletin, 28:270–274, (2003).

67. Zhang Q, Yang X, Li P, Huang G, Feng S, Shen C, Han B, Zhang X, Jin F, Xu F, Lu TJ. Bioinspired engineering of honeycomb structure – Using nature to inspire human innovation. Progress in Materials Science, 74:332–400, (2015).

68. Ha NS, Lu G. A review of recent research on bio-inspired structures and materials for energy absorption applications. Composites Part B: Engineering, 181:107496, (2020).

69. An X, Fan H. Hybrid design and energy absorption of luffa-sponge-like hierarchical cellular structures. Materials & Design, 106:247–257, (2016).

70. Tsang HH, Raza S. Impact energy absorption of bio-inspired tubular sections with structural hierarchy. Composite Structures, 195:199–210, (2018).

71. Tsang HH, Tse KM, Chan KY, Lu G, Lau KT. Energy absorption of muscle-inspired hierarchical structure: Experimental investigation. Composite Structures, 226:111250, (2019).

72. Yu X, Pan L, Chen J. Experimental and numerical study on the energy absorption abilities of trabecular–honeycomb biomimetic structures inspired by beetle elytra. Journal of Material Science, 54:2193–2204, (2019).

73. Zhang Y, Gao L, Xiao M. Maximizing natural frequencies of inhomogeneous cellular structures by kriging-assisted multiscale topology optimization. Computers & Structures, 230:106197, (2020).

74. Huang X, Zuo ZH, Xie YM. Evolutionary topological optimization of vibrating continuum structures for natural frequencies. Computers and Structures, 88:357–364, (2010).

75. Singiresu SR. Mechanical Vibrations, Sixth edition. Peasrson Education Limited, United Kingdom, (2018).

76. Qiu C, Guan Z, Jiang S, Li Z. A method of determining effective elastic properties of honeycomb cores based on equal strain energy. Chinese Journal of Aeronautics, 30:2, (2017).

77. Autar KK. Mechanics of Composite Materials. New York: Taylor & Francis, (2006).

78. Vapnik V. The Nature of Statistical Learning Theory. New York: Springer, (1995).

79. Rasmussen CE, Williams CKI. Gaussian Processes for Machine Learning. Cambridge, Massachusetts: MIT Press, (2006).

7 Design and Optimization of Mechanical Metamaterials Using Correlation Analysis

7.1 INTRODUCTION

In recent years, 3D printing or additive manufacturing (AM), a technology invented for prototyping in the early 1970s, has become a popular tool for manufacturing larger parts or engineering structures with complex geometries. It is widely believed that AM is the next-generation manufacturing tool and will change the paradigm of materials design and manufacturing [1,2]. In fiber-reinforced polymer composite structures, some high-profile new developments include printing sandwich cores with short fiber-reinforced thermosetting polymers [2] and printing continuous fiber-reinforced polymer composites [3]. A remarkable development in AM is 4D printing. Since the terminology of 4D printing was coined by Tibbits in 2013 [4,5], the subject has attracted considerable attention in both academia and industry [6,7]. Four-dimensional printing is defined as the time evolution of the 3D-printed structure, in terms of shape, property, and functionality [7,8]. A natural choice for 4D printing is to use shape memory polymer (SMP) as ink because SMP changes shape with time upon stimuli. Four-dimensional printing with SMPs is a very popular area of research, and many studies have been conducted in the past several years [9–39].

Several printing approaches have been utilized for 4D printing SMPs, including digital light processing, direct ink writing, extrusion, and recently volumetric printing [9, 18, 26, 40–51]. AM technologies are increasingly used to produce end-user, multicomponent, and multi-material parts for various applications. AM has been used in almost all types of materials and applications (polymer, metal, ceramic, tissue, composite, etc.). The literature is growing exponentially [52–62]. Although AM is a promising manufacturing technology, such layer-by-layer approaches limit quantity, degrade surface quality, constrain geometric capabilities, increase postprocessing requirements, and may cause anisotropy of mechanical performance. A manufacturing technique capable of simultaneously fabricating all points within an arbitrary 3D geometry would provide a different strategy to address these issues

DOI: 10.1201/9781003400165-7

and complement existing AM methods. A method that forms parts volumetrically allows for different ways to integrate multiple components and may widen the material landscape to enhance the functionality of finished parts. This is why volumetric 3D printing has emerged in the past several years. In volumetric 3D printing, an entire 3D object is simultaneously solidified by irradiating a volume of liquid photocurable resin from multiple angles with dynamic light patterns. Unlike most other AM methods, tomographic volumetric AM is layer-less, meaning that it does not fabricate objects by solidifying one voxel, one line, or one layer at a time. Instead, light from the subsequent tomographic patterns builds up an energy dose within the complete volume of the target object. In addition, support struts are not needed because the printed part is supported by surrounding liquid resin. Printing time, around a few seconds [62], and centimeter-scale prints with resolutions down to micrometer scales [61] make volumetric 3D printing a future-driven 3D printing method. More broadly, volumetric 3D printing is capable of rapid production of complex parts in a broad range of polymeric materials (e.g., high viscosity acrylates, thiol-enes, cell-laden hydrogels, preceramic, and glass-forming polymer resins [62]). The swift growth of volumetric 3D printing for myriad applications opens many opportunities [57, 63–67].

In structural applications, the combination of 4D printing and lightweight structures drives the paradigm shift in structural design. As compared to other 3D-printed polymeric structures, an obvious advantage of the SMP-based structures is that these structures can change shape upon stimuli and output useful mechanical energy, or they are able to do positive work to their surroundings, for example, as deplorable structures in aerospace applications [68–73], sutures, stents, scaffolds in medical and biomedical applications [74–81], crack-closing devices in damage self-healing applications based on the close-then-heal strategy [82–107], artificial muscles by twist insertion in SMP fibers [108–115], and many more. All these applications need the SMP structures or devices to overcome some types of constraints and do some positive work to the constraints so that the actuation can be activated and performed.

In order for the SMP devices or structures to output more mechanical energy, it depends on two critical issues: energy input and energy storage. Energy input is through mechanical deformation. The SMP devices or structures need to be deformed through external loading or pre-straining. Microscopically, mechanical deformation causes the alignment of molecular segments to be aligned along the loading direction, leading to the ordering of the network, or reduction in conformational entropy. After cooling, unloading, and some spring back, a temporary shape can be fixed, and the SMP network is in a nonequilibrium status with reduced conformational entropy. Therefore, the input energy is stored in terms of entropy reduction. This completes the programming stage. When temperature rises above the transition temperature (either glass transition or melting transition), entropy increases autonomously based on thermodynamics law. As a result, the SMP releases energy and, macroscopically, returns to its original shape. However, it has been proved that entropy reduction alone cannot store enough energy; as a result, programmed SMPs cannot do substantial work. The reason is that while entropy-driven SMPs can have excellent shape recovery, they have very limited recovery stress. The very low recovery stress makes it difficult to overcome constraints and to do positive work. Consequently, the recovery stress or the ability of the SMP to do positive mechanical work is a grand challenge.

To deal with this challenge, Fan and Li [116] proposed a new strategy to store mechanical energy, i.e., by enthalpy increases through bond length change and bond angle change. This strategy proves to work. Several SMPs or even self-healable and recyclable SMPs such as shape memory vitrimers have been designed, synthesized, and tested [117,118], which have exhibited record-high recovery stress. This strategy has also been proved by molecular dynamics simulation and solid mechanics modeling [119,120]. In addition to using enthalpy increase to store mechanical energy, cold programming, i.e., programming in a glassy state, has also been proved to result in higher recovery stress [121–124]. Because of the importance of increasing the recovery stress, numerous other studies have focused on investigating various techniques to enhance the recovery stress of SMPs, for example, using nanocomposites [125,126]. Figure 7.1 shows a schematic of using both entropy and enthalpy to store mechanical energy.

It is worth noting that most reported recovery stress values in the existing literature are derived from tests conducted on solid samples such as cubes or cylinders. It is widely recognized that the specific load-carrying capacity of a material can be significantly enhanced when it is manufactured into various structures such as I-beams, T-beams, box beams, and sandwich beams, rather than remaining in the form of solid cuboid beams. Therefore, it is expected that by 3D printing

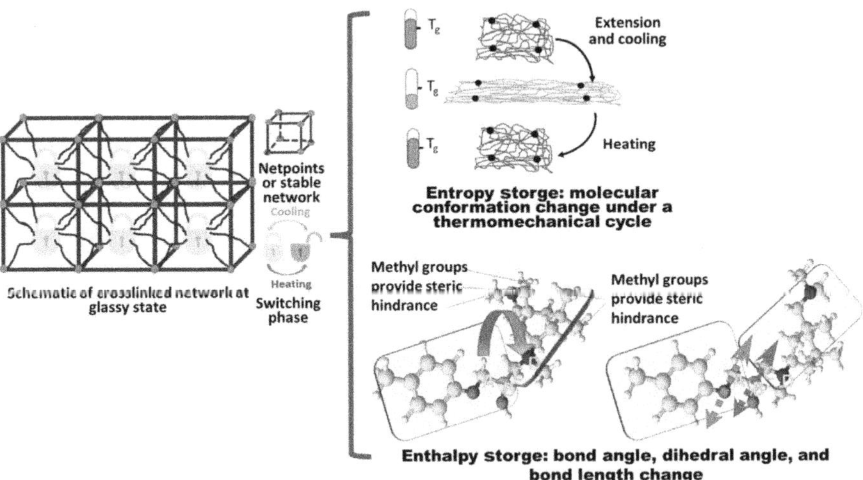

FIGURE 7.1 Schematic of the mechanism for a shape memory polymer (SMP) network to store mechanical energy. The necessary condition for the polymer to have a shape memory effect is that the polymer must have a stable network and a switching phase; to be sufficient, the polymer must have a mechanism for energy storage, here through both entropy reduction and enthalpy increase.

SMPs into metamaterials, the specific recovery stress can be enhanced compared to solid SMP structures.

Recently, SMPs have been used as ink in 4D-printed metamaterials [27, 127–133]. Lattice unit cells and lattice structures have been printed and tested [27, 127]. A metamaterial design paradigm using gears with encoded stiffness gradients as the constituent elements and organizing gear clusters for versatile functionalities was conducted [128]. Auxetic structures with adjustable mechanical properties were developed and fabricated through the utilization of SMPs, offering potential applications in the field of medical devices [129]. In order to create a hierarchically structured metamaterial with strain-dependent solid–solid phase transformation, researchers designed and manufactured a material with tunable mechanical properties, suitable for micro-actuators, grippers, and programmable devices [130]. Various two- and three-dimensional auxetic structures have been proposed, considering parameters such as mass, buckling load, natural frequency, Poisson's ratio, and compression strength and comparing them to extensive numerical and experimental results [131]. These auxetic structures exhibit unique characteristics based on their structural orientation, leading to potential applications in the medical, sports, and automotive industries [132]. Topology optimization methods have predominantly been employed to optimize the performance of these metamaterials, aiming to achieve maximum efficiency [133]. However, it should be noted that a limitation of topology optimization lies in its limited capability to optimize structures based on a single-parent design despite the presence of a vast design space for potential global optima.

This chapter introduces a straightforward design criterion aimed at identifying optimized metamaterials that exhibit exceptional shape memory properties, particularly, recovery stress. Achieving optimal shape memory performance necessitates a structure that is both flexible, allowing for substantial displacements during programming, and strong, enabling the storage of higher programming stress and energy. To reconcile these contrasting requirements within lightweight structures, we initially focused on bending-dominated designs. It has been demonstrated that bending-dominated structures possess remarkable flexibility [134,135], satisfying the need for larger displacements. Hexagonal honeycomb structures with a periodic arrangement were investigated due to their high shear strength and shear strain in the in-plane direction, making them suitable for applications involving passive morphing airfoils [136]. By optimizing bending-dominated structures to possess superior strength or greater programming stress, the second requirement is fulfilled. Consequently, bending-dominated structures, which ensure higher displacement and higher deformation, if they are also endowed with high strength, should have the potential to input more energy and store more energy and thus should exhibit superior recovery stress.

To address these objectives, we investigate the design spaces of 3D bending-dominated lattice unit cells and thin-walled cellular unit cells to identify structures with exceptional strength and recovery stress properties. While stretch-dominated lattice structures and thin-walled cellular structures with extremely low relative density can exhibit significant deformation through buckling, they do not meet the requirement for high programming load and are therefore not considered in this study.

While in Chapter 6 several thin-walled unit cells with superior natural frequency were explored, there remains a research gap in exploring a wide range of design space. Although techniques like topology optimization can lead to optimal structures, a data-driven approach that is closer to the global optima can be achieved through supervised machine learning regression models and statistical analysis techniques. In this chapter, a novel design optimization framework that utilizes existing regression models and statistical analysis to discover novel optimized thin-walled cellular structures that possess superior strength and recovery stress properties is presented. While the individual models used in this chapter have been widely employed in various research, the combination of these techniques for structural optimization has not been proposed before.

For comparison purposes, the optimal 3D lattice unit cells are adopted from our previous studies in Chapter 6, where machine learning regression models and generative adversarial networks (GANs) were employed to optimize them in terms of uniaxial and multiaxial strengths. The reason is that to have higher recovery stress, the unit cell must have high compression strength, which is the criterion used in Chapter 6 in discovering new lattice unit cells. Although we also discovered several new thin-walled cellular unit cells, the criterion used is higher natural frequency. Therefore, these cellular unit cells do not necessarily exhibit high recovery stress. Therefore, in this chapter, we need to first discover new thin-walled cellular unit cells with high compression strength using an inverse machine learning framework. In this chapter, a new inverse design framework based on correlation analysis, instead of GAN, is established and presented. Maxwell's criterion for rigidity of frames is employed to select optimal bending-dominated lattice unit cells. This criterion, with a few assumptions, is then extended to thin-walled cellular structures for unit cell classification and selection. Numerical simulations and experimental comparisons are conducted to validate the proposed inverse design framework and optimal structures. Both bending-dominated lattice unit cells and thin-walled cellular unit cells exhibit high strength, good flexibility, and record-breaking recovery stress.

7.2 DATA GENERATION AND FINGERPRINTING OF THIN-WALLED CELLULAR UNIT CELLS

To study the thin-walled structural performance in the in-plane orientation and optimize them for superior stress recovery properties, the following steps are followed. First, to explore a larger space of structural design, a dataset of possible thin-walled cellular structures within a design space shall be formed using a representative volume element (RVE). Second, machine learning regression models or forward design models using a training dataset that can predict the structural properties of any unit cell within the RVE shall be developed. The machine learning regression models assist in drastically reducing the property prediction time and computational power. Third, a design criterion to differentiate the unit cells based on their structural behavior (i.e., bending- or stretching-dominated) shall be proposed. Finally, an inverse design framework using machine learning regression models and statistical analysis tools shall be assembled to predict novel lightweight structures with superior strength and recovery stress.

To explore a wide range of novel structural designs, certain boundary conditions should be set such that a large variety of unexplored design spaces with extensive structural properties can be covered without making the optimization task too complex. For this purpose, an RVE with nine points (joints in 3D) is considered as shown in Figure 7.2. A total of 20 lines (thin walls in 3D) can be formed by connecting every two adjacent points. Here, for simplicity, only the nearest neighboring points are connected to form a line. For example, points 1 and 2 can be connected to form the line (12), but 1 and 3 cannot be connected as there is a point (point 2) in between them. Similarly, with various combinations of these lines and by mirroring the RVE into the horizontal and vertical axis for each combination, a huge dataset of nearly a million different thin-walled structures can be obtained. It is noted that if we allow the connection of non-nearest neighboring points such as 18 and 19, a total of 35 lines will be created, which will lead to a much larger training dataset and may capture more optimized structures. For simplicity and to demonstrate the machine learning framework only, we used 20 lines in this study.

To train a machine learning algorithm (supervised), a training dataset containing both the inputs, which are the structures, and the outputs, which are the

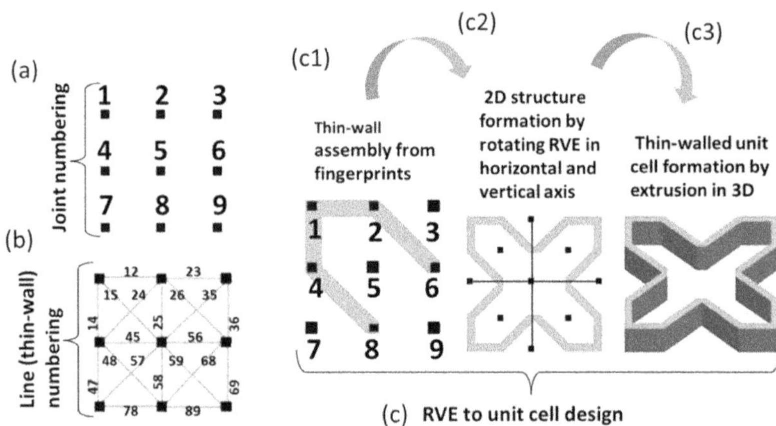

FIGURE 7.2 Representative volume element (RVE) for the thin-walled unit cells. (a) The RVE consists of 9 joints that can be used to form a total of 20 lines (thin walls). (b) By connecting the neighboring points, a total of 20 lines (thin walls) can be formed. Any of the two numbers in a line is a fingerprint of that line. (c) Procedure from the RVE to unit cell: (c1) collect the fingerprints of the lines in (b) in vector format to form a quarter of a unit cell, for example, (12 14 26 48); (c2) mirror the quarter unit cell in vertical and horizontal axes to form a symmetric 2D unit cell; and (c3) extrude the 2D unit cell in the out-of-plane direction to create a 3D unit cell.

desired mechanical properties of each structure, should be provided. The structures must be fingerprinted for the machine learning algorithm to interpret the data. Fingerprinting is the process of converting each structure into a machine-readable logical sequence or pattern of digits. In this study, each structure is initially named with the combination of all the joints forming that particular unit cell within the RVE. For example, the thin-walled structure in Figure 7.2 is fingerprinted as (12 14 26 48), where (12) represents the line or wall formed by connecting points (joints) 1 and 2. Similarly, (14), (26), and (48) represent lines or walls connecting points 1 and 4, 2 and 6, and 4 and 8, respectively. Since the position of the joints and their digital representations are fixed, each structure formed within the RVE can be given a unique logical fingerprint. This procedure can be easily followed to convert any structure into a fingerprint and *vice versa*. For the machine learning regression model training, these fingerprints are further converted into a binary format which improves the prediction accuracy. This is done by assigning a fixed position for each formed line in a vector and by representing all the line positions present in a particular structure with 1s and the rest as 0s in the vector. In this study, line 12 is assigned the 1st position in the binary format vector, and similarly, 14, 26, and 48 are given the 7th, 15th, and 17th positions in the vector. Hence, the fingerprint of the cellular structure in Figure 7.2 will be represented in the binary format as (1 0 0 0 0 0 1 0 0 0 0 0 0 0 1 0 1 0 0 0). While all the unit cells formed using the proposed RVE will be unique to one another, some structures will be repetitive if tessellated to form an infinite lattice. For example, unit cells (15 59 89 78 47 14) and (35 57 78 89 69 36) will form the same lattice structures when tessellated. The two unit cells are shown in Figure 7.3. Since this study is focused on optimizing the unit cells only, the repetitive fingerprints (if tessellated to form an infinite lattice) were ignored. This facilitates the ease of data handling and data generating processes. However, if one is to use the RVE to create lattice structures by tessellation, attention must be paid to the possibility that different unit cells may create the same lattice structure. For example, for the two unit cells in Figure 7.4, although they are different unit cells at the unit cell level, they create the same cellular structures after tessellation.

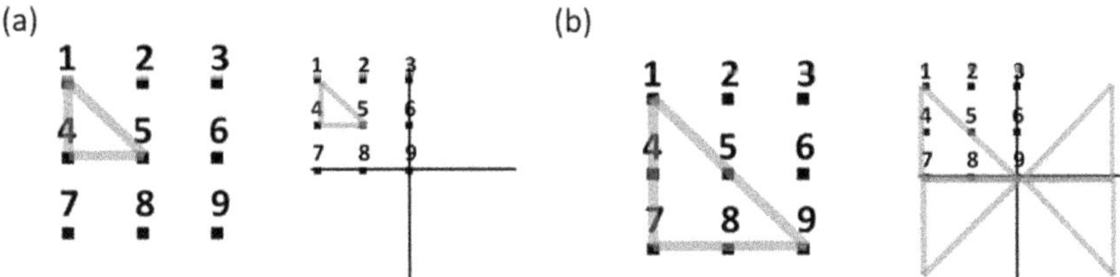

FIGURE 7.3 Sample thin-walled cellular unit cell generation and fingerprinting method. (a) Fingerprint (15 14 45) and (b) fingerprint (15 59 89 78 47 14). From this figure, it is seen that fingerprint (a) cannot form a cellular unit cell as there are missing connections after mirroring along the vertical and horizontal axes, while fingerprint (b) can form a meaningful cellular unit cell because after mirroring along the vertical and horizontal axes, a unit cell with all the members connected is formed.

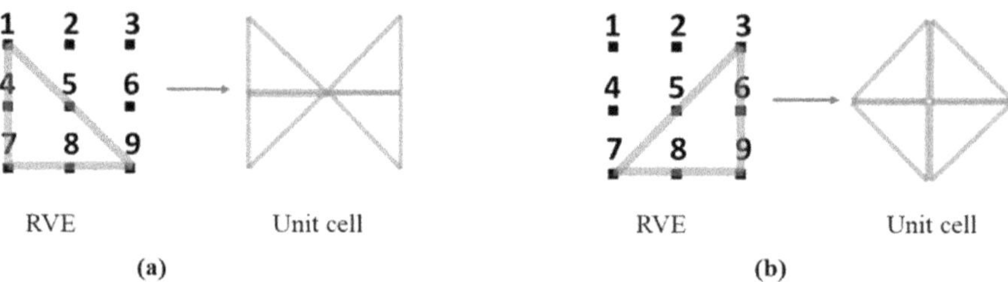

FIGURE 7.4　(a) Fingerprint (15 59 89 78 47 14) and (b) fingerprint (35 57 78 89 69 36). While they are different unit cells, they form the same lattice structure when they are tessellated to form an infinite lattice structure.

In this study, all the structures are fingerprinted using the procedure stated earlier. MATLAB combination function is used to generate a dataset of all possible structures within the RVE using simple coding. A training dataset of 2000 fingerprints is extracted randomly from the untrained dataset for the machine learning regression analysis. MATLAB functions such as "y = datasample(data,k)" are used for the training dataset extraction.

7.3　FORWARD MACHINE LEARNING REGRESSION MODEL FOR THIN-WALLED CELLULAR UNIT CELLS

The main motivation for this study is to predict optimal thin-walled structures with superior recovery stress. As discussed earlier, directly calibrating the shape memory effect or the recovery stress of these structures is a complex and time-consuming procedure, both experimentally and numerically [137]. The experimental analysis involves structure manufacturing, experimental setup, and a multiphase shape memory training, which is extremely time-consuming, especially when multiple samples are involved. The numerical analysis can also be very complex as it involves many curve-fitting parameters, nonlinear material properties, and thermomechanical analysis [138]. Since the training dataset is comparatively large, it is impossible to adopt any of the above conventional procedures.

For this purpose, as this is a structural optimization problem if the material properties, overall volume, and test boundary conditions for all the structures are kept the same, the overall recovery stress of a structure will depend on the energy stored during programming and the total strain stored. For a given SMP, its recovery stress depends on the programming strain only because the shape fixity ratio, shape recovery ratio, and rubbery modulus are SMP-dependent only. In other words, higher displacement will lead to higher strain and thus higher energy storage and better recovery stress [121]. The maximum displacement occurs when the structure or a structure element fails. Therefore, the higher displacement requirement can be translated into higher mechanical strength. Hence, under uniaxial compression, the compression strengths of all the structures in the training dataset are recorded along with their masses. ANSYS simulation software is used to model (workbench

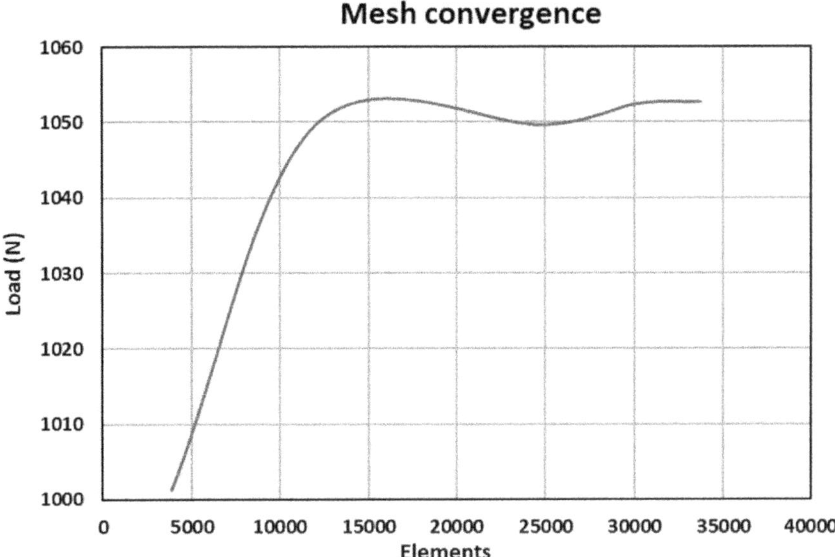

FIGURE 7.5   Mesh convergence analysis.

design modeler) and evaluate all the 2000 thin-walled cellular unit cells. The data generation is performed on a workstation with 32 GB RAM and an i7 processor, and the total computational time for the training dataset is about 75–85 man-hours. Mesh convergence analysis is conducted for consistent results; see Figure 7.5. In this study, 74163 nodes and 10422 elements were used to model the thin-walled cellular unit cells. The numerical analysis is performed by considering only the elastic properties of the base material to minimize complexities and time consumption that might rise if the viscoelastic properties are considered. Once a material is selected, the recovery stress of the structures depends on their programmable strain which is governed by the number of elements and their orientations. Therefore, the model should be applicable irrespective of the material properties. Once the training dataset is ready with the input fingerprints and the output mass and compression strength properties, the MATLAB regression analysis tool is used to compare several machine learning algorithms for their prediction accuracies with the training dataset. With a fivefold validation, the Gaussian process regression (GPR) model outperformed other machine learning models such as ensemble tree and support vector machines (SVMs with a root-mean-square error (RMSE) less than 5% and R-squared value of 0.98 for both the mass and compression strength properties; see Figure 7.6. The model parameters are given in Table 7.1, and model comparisons are summarized in Table 7.2. The GPR model is a kernel-based probabilistic model which uses a set of random variables having a Gaussian distribution to do the predictions. Previously, the GPR models were proven to work best compared to other models for supervised machine learning regression models, especially with structural data and their mechanical property predictions.

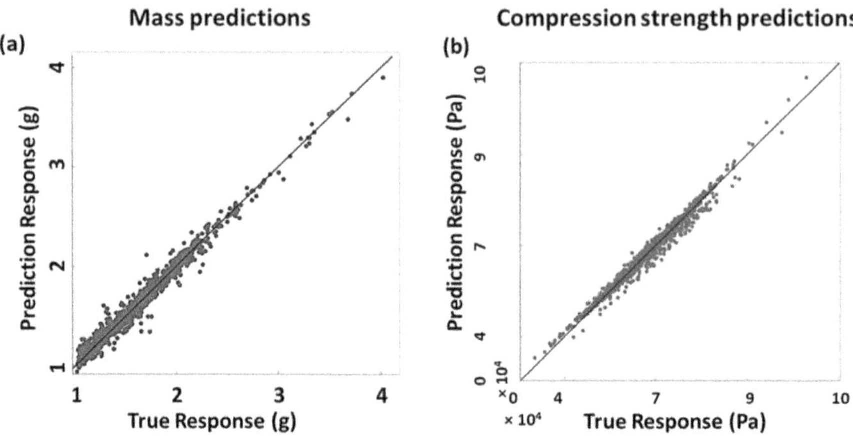

FIGURE 7.6 Gaussian process regression (GPR) models for (a) mass and (b) compressive strength predictions. A training dataset of 2000 fingerprints is used to train the GPR model, and the prediction accuracy R-squared value is 0.98. The blue dots (•) represent the observations, and the inclined solid lines represent perfect predictions. The roughly even scattering of the observation along the perfect predictions represents a solid regression model.

TABLE 7.1

Machine Learning Model Parameters

Model Type	Gaussian Process Regression (GPR)
Basic function	Constant
Kernel function	Matern 5/2
Root-mean-square error (RMSE)	Mass: 0.0564, compression strength: 0.02084
Mean absolute error (MAE)	Mass: 0.08, compression strength: 0.014485
Prediction speed	~5700 obs/sec
Training time	503.23 sec

7.4 SELECTION CRITERION FOR OPTIMAL STRUCTURES

Now, we have the forward design model for the thin-walled cellular unit cells; together with the forward design model in Chapter 5 for lattice unit cells, we can discover new unit cells with high recovery stress. Therefore, the following discussions are required for both types of unit cells, meaning lattice unit cells and cellular unit cells.

We need to first answer a question: how to measure the recovery stress. Or which parameters directly or indirectly determine the recovery stress? Based on our previous discussion, in order to have higher recovery stress, we need to have higher energy input and higher energy storage. Another parameter is the shape recovery ratio, i.e., we need a higher shape recovery ratio. Therefore, we first need the unit cell to have higher mechanical strength so that more mechanical energy

TABLE 7.2

Gaussian Process Regression (GPR) Model for Mass and Compression Strength Predictions

Uniaxial Compression Strength

Machine Learning Technique	Training Dataset			Testing Dataset		
	Root-Mean-Square Error (RMSE)	R^2	Mean Absolute Error (MAE)	RMSE	R^2	MAE
Matern 5/2 GPR	0.02084	0.98	0.014485	0.021457	NA	0.0189
Ensemble (bagged tree)	0.02488	0.92	0.010154	0.028471	NA	0.0154
Cubic support vector machine (SVM)	0.02321	0.89	0.01356	0.035481	NA	0.0245

Mass

Machine Learning Technique	Training Dataset			Testing Dataset		
	RMSE	R^2	MAE	RMSE	R^2	MAE
Matern 5/2 GPR	0.0564	0.98	0.08	0.045	NA	0.084
Ensemble (bagged tree)	0.0721	0.96	0.1455	0.084	NA	0.1247
Cubic SVM	0.0785	0.92	0.1487	0.041	NA	0.14

can be input. The compressive strength of any structure within the specified design space can be calibrated using the regression or forward design model. The energy input to a unit cell depends on both the mechanical strength, which can be translated to force, and displacement. Therefore, the second requisite, which is the maximum displacement of a unit cell before failure, is also crucial for optimal recovery stress. For this purpose, the study of lattice structures and their behavior based on the number of elements and joints shall be referred to. In general, when a lattice structure is under axial loading, the elements or rods carry the load. Based on Maxwell's criterion for rigidity of frames (Equations 7.1 and 7.2), the lattice structures can be either stretching- or bending-dominated [139–142]. It is noted that Maxwell's criterion was originally for pin-jointed structures. Three-dimensional printed joints are, generally speaking, not pin joints. Furthermore, there are well-known exceptions to the criterion, and buckling impugns Ashby's original intent in using the criterion. Recently, some researchers have extended Maxwell's criterion to 3D-printed structures that have elastic or rigid or frozen joints [143–146]. In this study, we used Maxwell's criterion as a rough guide to help us select bending- or stretching-dominated unit cells.

$$M = b - 2j + 3, \text{ 2D structures (frames)}, \tag{7.1}$$

$$M = b - 3j + 6, \text{ 3D structures}, \tag{7.2}$$

where b is the number of truss members and j is the number of frictionless joints. Here, if $M \geq 0$, the structure is stretching-dominated, and if $M < 0$, the structure is bending-dominated.

In stretching-dominated structures, the mode of failure is due to the rod stretching or buckling, while the bending-dominated structures fail primarily due to the rod bending. Thus far, several reports have suggested that stretching-dominated structures have higher strength and toughness as compared to their bending-dominated counterparts due to the rigidity in the framework [139–142]. In our previous studies [147], we reported several novel lattice unit cells with superior compression strength properties compared to the classical octet lattice unit cell. However, we did not print the structures with SMP and did not explore their recovery stress properties. On further investigation, contrary to the literature up to now, it is observed that several of the optimal structures, which are bending-dominated, exhibit similar or even higher relative compression strength compared to their stretching-dominated counterparts. In our previous study [147], we considered an RVE with 27 joints and 162 truss elements. Among the 550 total orthotropic lattice unit cells that can be generated within this RVE, the optimal bending-dominated lattice unit cells are observed to outperform any other stretching-dominated lattice unit cells, including the octet unit cell, under uniaxial compression. More details on the RVE and comparisons of several optimal lattice unit cells with respect to mass can be found in our previous study [147]. Numerical and experimental comparisons with bending-dominated lattice structures can be found in the section for validation. These optimal bending-dominated unit cells can be very advantageous to serve as shape memory structures due to their strong, lightweight, and flexible bending responses. The undesired buckling phenomenon of the lattice elements can be avoided by replacing the standard cylindrical elements with biomimetic rods that have higher buckling strengths [148]. From Ref. [147], we selected different optimal bending-dominated lattice unit cells to compare them with the stretching-dominated octet lattice structure. Figure 7.7

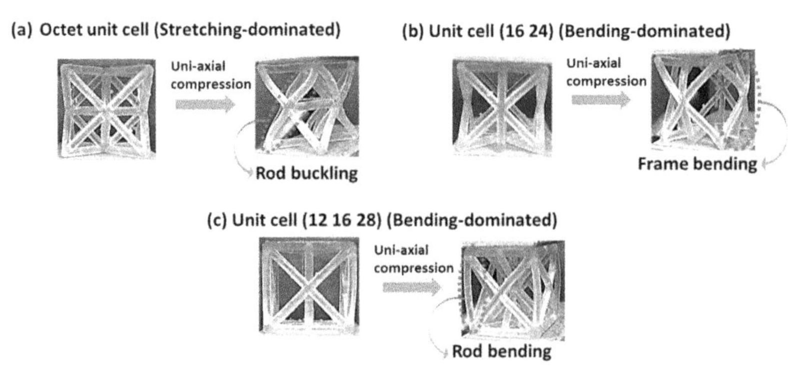

FIGURE 7.7 Behavior of stretching- and bending-dominated lattice unit cells under uniaxial compression. (a) Octet unit cell, (b) unit cell (16 24), and (c) unit cell (12 16 28). The octet unit cell can be observed to display a stretching-dominated behavior with local rod buckling, but the proposed bending-dominated unit cells display a global or local rod bending behavior. It is noted that during the test, if bending occurs suddenly, we define it as buckling. If bending grows with load, we define it as a bending-dominated structure.

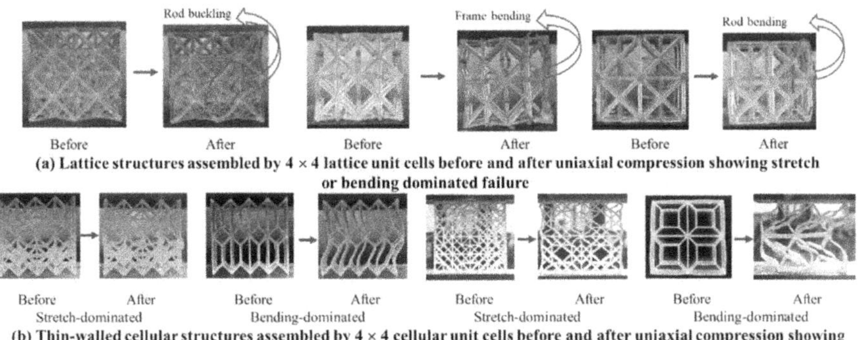

(a) Lattice structures assembled by 4 × 4 lattice unit cells before and after uniaxial compression showing stretch or bending dominated failure

(b) Thin-walled cellular structures assembled by 4 × 4 cellular unit cells before and after uniaxial compression showing stretch or bending dominated failure

FIGURE 7.8 (a) Lattice structures assembled by 4 × 4 lattice unit cells exhibiting stretching- and bending-dominated failure modes under uniaxial compression and (b) thin-walled cellular structures assembled by 4 × 4 cellular unit cells exhibiting stretching and bending-dominated failure modes under uniaxial compression.

presents the 3D-printed unit cells before (left) and after (right) deformation due to uniaxial compression. Figure 7.8 shows that the unit cells can be assembled into a larger lattice by 4 × 4 unit cells.

In the case of the thin-walled cellular unit cells, there is no classification criteria so far, and Maxwell's criterion was not applied due to the local bending and buckling behavior of the thin walls, especially in the out-of-plane orientation. But in the in-plane orientation, although the local bending or buckling of the walls is predominant, the global thin-walled unit cell might display a 2D frame-like behavior. To substantiate this, we applied the 2D Maxwell's criterion for rigidity of frames (Equation 7.1) to design structures distinguished as bending- and stretching-dominated thin-walled cellular unit cells by ignoring the local bending and buckling modes. Using stereolithographic AM (Formlabs, Form 3 system) and a commercial polymer (Clear), we printed these unit cells (20 mm height and 1.5 mm wall thickness) to observe their behavior under uniaxial compression tests. The polymer used has a density of 1.16 g/cm^3, compressive strength of 201 MPa, compressive yield strength of 65.9 MPa, and modulus of elasticity of 2600 MPa. A test speed of 0.5 mm/min and uniform cell wall thickness are maintained for all the structures. The structure in Figure 7.9(a) is classified as stretching-dominated as M is greater than zero based on the number of wall elements ($b = 40$) and the number of joints ($j = 21$), and the structure in Figure 7.9(b) is classified as bending-dominated as M is less than 0 ($b = 24$, $j = 21$). Note that the structures in Figure 7.9(a) are not optimized and only used to depict the structural behavior. Similar behavior can be seen in 4 × 4 cellular structures in Figure 7.8(b).

From Figure 7.9(a), the structure classified as a stretching-dominated thin-walled structure exhibits a global stretching-like behavior by fracturing at the peak load, while the bending-dominated structure (Figure 7.9(b)) exhibits global and local bending behavior. It can also be observed from Figure 7.9(b) that the bending-dominated structure lasted much longer and had higher displacement

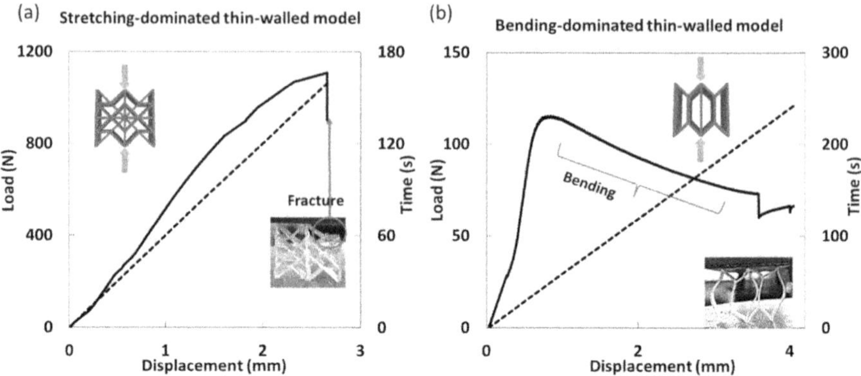

FIGURE 7.9 Experimental load vs. displacement (solid curve) comparisons over time (dotted line) under uniaxial compression for (a) stretching-dominated thin-walled cellular unit cell and (b) bending-dominated thin-walled cellular unit cell. The solid lines represent load vs. displacement, and the dotted lines represent displacement vs. time comparisons. The bending-dominated unit cell (b) can be observed to show lower load-bearing capacities, higher displacements, and flexible deformation behavior in the in-plane direction compared to the stretching-dominated unit cell (a). It can also be seen that the unit cell in (b) requires a longer time before failure (300 sec), supporting the bending-like behavior, and the unit cell in (a) requires less time before failure, representing stretching-like behavior. Furthermore, it is seen in (a) that the load dropped suddenly, suggesting fracture of some rods, while the load drops gradually in (b), suggesting bending.

but lower load-carrying capacity compared to the stretching-dominated structure. Similar observation can be made for cellular structures made of 4×4 cellular unit cells, as shown in Figure 7.8(b). Therefore, the bending-dominated structures satisfy the requirement for larger displacement. However, as mentioned earlier, the two structures in Figure 7.9 are not optimized. They are used for demonstration purposes only. In this study, we will optimize the bending-dominated structures so that they will also have higher strength, in addition to higher deformability, which may lead to higher recovery stress. Based on these experimental observations, it can be presumed that the 2D Maxwell's criterion for frames can be extended into thin-walled unit cells in the in-plane orientation to classify thin-walled structures as either bending- or stretching-dominated structures. Since Maxwell's criterion was originally proposed for pin-joined structures, the following assumptions were made to apply this criterion for thin-walled structures in the in-plane orientation: (1) Local bending or bucking of the thin walls should be ignored. (2) The criterion should be used just as a preliminary screening approach to assess the overall structural behavior in the in-plane orientation only. Also, it should be noted that local buckling in rods is observed in bending-dominated structures as well and this buckling is caused due to overall structural bending. The buckling of rods in stretching-dominated structures, however, is due to the overall structural stretching-like behavior.

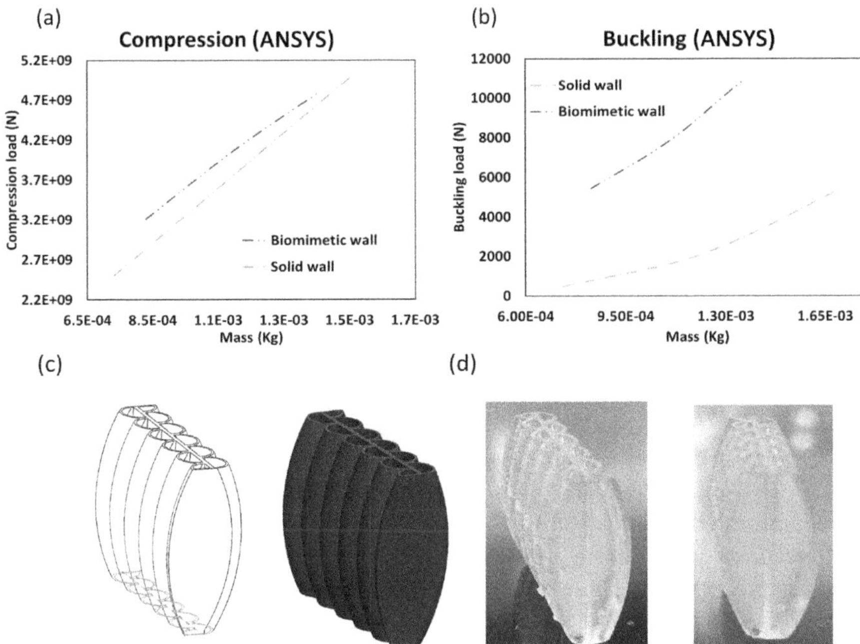

FIGURE 7.10 Comparisons of biomimetic wall inspired from gain clam seashell for (a) compression load, (b) buckling load, (c) biomimetic designs, and (d) 3D-printed samples.

Now, by using this criterion and designing bending-dominated structures with higher load-carrying capacities (compression strength), optimal thin-walled unit cells that are strong as well as flexible can be created. These unit cells may lead to multifunctional applications such as high strength and high recovery stress. To mitigate the local buckling of the thin walls, we designed hybrid wall structures by mimicking giant clam seashell structures. To demonstrate that biomimetic walls can improve the buckling resistance, we designed gain clam seashell-inspired walls and conducted finite element simulation. The simulation results between the solid wall and biomimetic wall are presented in Figure 7.10(a) and (b) for compression and buckling, respectively. Figure 7.10(c) and (d) shows the biomimetic wall design and 3D-printed samples, respectively.

7.5   CORRELATION ANALYSIS

In Chapter 6, we presented inverse machine learning frameworks for discovering lattice unit cells and thin-walled cellular unit cells with superior mechanical properties such as high compression strength, high natural frequency, and high-impact energy absorption. The inverse design framework is a combination of GANs and machine learning regression models. A GAN consists of two neural networks: a generator and a discriminator. The duty of the generator is to generate data and feed it to the discriminator. The discriminator, which is fed with a training dataset (unit cell fingerprints in this context), is trained to distinguish the data from the generator into real or fake data. The real data are fingerprints similar to the training data but not exactly the same, and the fake data are random noise. Now both the generator and discriminator run in a loop to train each other (based on the training dataset fed to the discriminator) until the generator learns to only generate real data (true fingerprints). While the machine learning regression or forward design model facilitates easier and faster property predictions, the GANs generate novel structures closer to the global optima. Though the GANs are proven to produce impressive results, they can involve extensive coding and training processes. Moreover, the inverse design framework using GANs developed in Refs. [149,150] needs to be run iteratively to optimize a set of inputs, leading to multiple optimization steps, and would produce a set of desirable designs rather than finding the most optimal solution within the design space. In this chapter, we present a new inverse design framework based on statistical tools like correlation analysis, which can greatly reduce the complexity of the inverse design problem and get much closer to the global optima.

Correlation analysis is a powerful statistical technique used to investigate the relationships between variables. It enables researchers to explore whether there is a meaningful association between different factors and to what extent changes in one variable correspond to changes in another variable. By conducting correlation analysis, researchers can gain valuable insights into the interdependencies within their data, identify patterns, and understand how variables influence each other. In traditional correlation analysis, one commonly used measure is Pearson's correlation coefficient. This coefficient, denoted as "r," assesses the strength and direction of the linear relationship between continuous variables. It ranges from -1 to 1, where -1 represents a perfect negative correlation (when one variable increases, the other decreases with a constant ratio), 1 indicates a perfect positive correlation (both variables increase or decrease together in a consistent manner), and 0 suggests no linear relationship between the variables.

However, when dealing with categorical data, such as variables with nominal or ordinal scales, Pearson's correlation coefficient is not appropriate since it assumes a linear relationship. Instead, alternative correlation measures are employed to evaluate the association between categorical variables. One widely used measure for analyzing the association between two nominal variables is Cramer's V. This measure ranges from 0 to 1, where 0 indicates no association and 1 signifies a perfect association. Cramer's V is often applied when working with contingency tables, which display the frequencies or proportions of the joint occurrence of the categories in the two variables. Another correlation measure suitable for categorical data is the phi coefficient (φ). It is used to assess the association between two dichotomous variables, which are variables with only two categories. The phi coefficient ranges from -1 to 1, where -1 represents a perfect negative association, 1 indicates a perfect positive association, and 0 suggests no association. The phi coefficient evaluates the relationship between variables based on the frequency of agreement or disagreement in their respective categories.

For analyzing ordinal variables, which have ordered categories, Kendall's Tau-b and Spearman's rank correlation are commonly employed. Kendall's Tau-b takes into account the rank order of the categories, allowing for the assessment of association between variables even when the specific values are not numerically meaningful. Spearman's rank correlation, similar to Kendall's Tau-b, focuses on the rank order but places more emphasis on the magnitude of differences between ranks.

By utilizing these correlation measures for categorical data, researchers can explore the relationships within their dataset and uncover associations between variables. This enables them to draw meaningful conclusions about the interdependencies and patterns that exist among categorical variables. The analysis of categorical data using correlation measures enhances the understanding of how different variables influence each other and provides valuable insights into the underlying connections and dependencies in the dataset.

When discussing the best categorical correlation analysis for assessing the strength of relationships between variables in multivariable data, Cramer's V coefficient is widely recognized as a suitable measure. This coefficient, which extends Pearson's chi-square statistic, takes into account the number of categories and the sample size of the variables under investigation [151]. Cramer's V is often preferred in multivariable data analysis due to its ability to provide a standardized measure of association that can be compared across different studies and datasets. It ranges from 0 to 1, where 0 indicates no association and 1 represents a perfect association, thus quantifying the strength and magnitude of the relationship between categorical variables. To compute Cramer's V, the chi-square statistic is normalized by dividing it by the minimum of $(n - 1)$ and $(r - 1)$, where n represents the sample size and r denotes the number of categories in the variable with the larger number of categories. This normalization accounts for variations in sample size and the number of categories, enabling fair comparisons [151]. Cramer's V offers several advantages when analyzing multivariable data. It provides a straightforward and interpretable measure of association, enabling researchers to assess the strength of relationships among multiple categorical variables within a single analysis. Additionally, it can be easily calculated using statistical software or by manual computation. Cramer's V coefficient is widely acknowledged as a suitable measure for evaluating the strength of relationships between categorical variables in multivariable data. It offers a standardized measure that facilitates comparisons across studies and datasets, allowing researchers to gain insights into the associations among multiple variables. However, it is important to note that Cramer's V solely measures the strength of association between variables and does not establish causality. It serves as a descriptive measure, aiding in quantifying the extent of the relationship between categorical variables.

It is noted that correlation analysis has been widely used in practice. Correlation analysis is a statistical method that measures the relation between an independent variable and a dependent variable. Techniques like Spearman correlation analysis have been used previously to investigate the relation between mechanical properties and mineral elemental content in shale (clastic sedimentary rock) [152]. Pearson correlation analysis (linear) has been used to evaluate the influence of the material variables and corresponded to the experimental result of fiber-reinforced cementitious materials [153]. Correlation analysis has also been extensively used in the healthcare industry such as to find the factors from a sample patient dataset that highly influence the emergency ward utilization with 88% prediction accuracy [154]. A high correlation deduces a strong influence of the independent variable on the dependent variable. The correlation analysis in this study is used to find the key elements in a structure which are the independent variables that could influence the desired mechanical property (compression strength) which is the dependent variable. Spearman correlation which is a monotonic analysis is employed for this study due to the nonlinear data type. Spearman correlation analysis works by considering the nonparametric measure of correlation between ranks of the two variables [154–156]. The Spearman correlation coefficient (r_s) is defined as follows:

$$r_s = 1 - \frac{6 \sum d_i^2}{N\left(N^2 - 1\right)},$$
(7.3)

where d is the difference between the two ranks of each observation and N is the number of observations.

7.6 INVERSE DESIGN FRAMEWORK USING CORRELATION ANALYSIS

The inverse design framework for this study is constructed by combining the machine learning regression models for the unit cell property prediction and correlation analysis to design novel optimal structures as shown in Figure 7.11.

To implement correlation analysis for the inverse design framework, 10 sub-datasets of 100 random fingerprints with their masses and compressive strengths are extracted. Correlation analysis is conducted on all the sub-datasets to find the elements (individual thin walls) that have a higher influence on the relative compression strength. Those elements having the highest influence (i.e., r_s close to 1) can be selected to form novel thin-walled structures. This will narrow down the optimization process and get much closer and faster to the global optima compared to GANs which would rather suggest multiple localized optimal suggestions for several iterations. The implementation of the correlation analysis on multiple subsets (10) is to validate the prediction accuracy of the framework. The final output will be novel lightweight thin-walled cellular unit cells with superior strength and flexibility, leading to higher recovery stress.

In this study, the design criterion is to use the correlation analysis to predict bending-dominated orthotropic thin-walled cellular unit cells with the highest specific strength possible within the RVE. Maxwell's criterion (Equation 7.1) is used to extract bending-dominated structures, and the forward machine learning regression models are used to predict the strength and mass properties of the

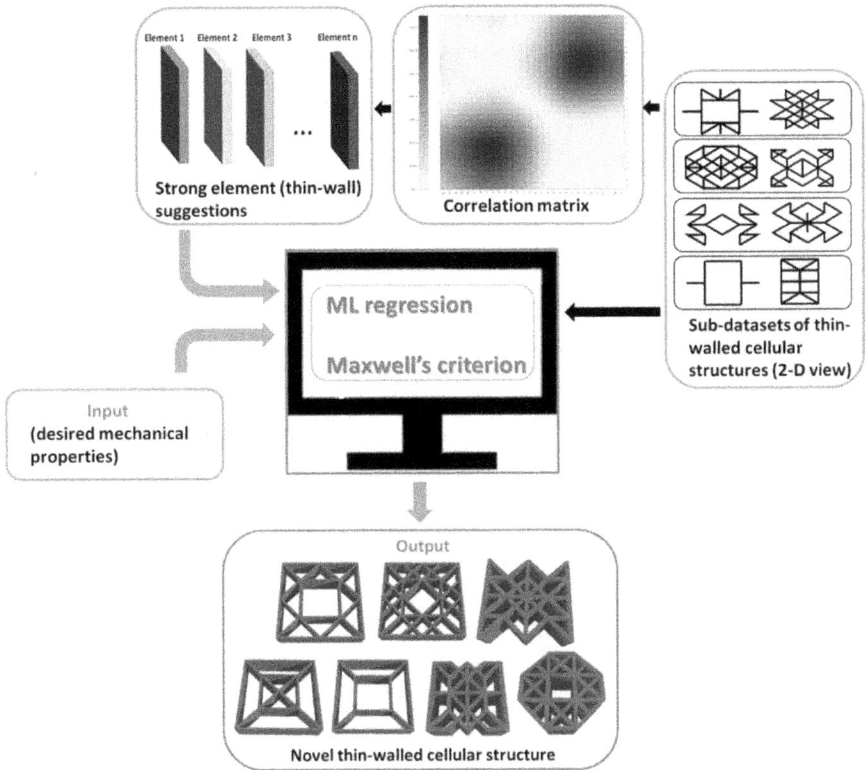

FIGURE 7.11 Inverse design framework for thin-walled cellular unit cells and cellular structures. The inverse design framework is formed by combining the ML regression models, correlation analysis, and the selection criterion. First, the machine learning regression models were trained using a training dataset to predict the mechanical properties of the thin-walled unit cells. The inverse design framework is later formed by applying correlation analysis to generate new designs with optimal strength and then extracting flexible structures by applying Maxwell's criterion. Hence, the input for the framework will be the desired mechanical properties, and the output will be optimal flexible and strong thin-walled cellular unit cells.

designs. The flowchart and steps to implement the inverse design framework are shown in Figure 7.12. The validation results using Maxwell's criteria are given in Table 7.2. It can be comprehended from the RVE that to extract structures with orthotropic symmetry, the total number of elements condenses to 12. Upon conducting correlation analysis on 10 subsets, each holding 100 different fingerprints and following the previously mentioned optimization framework, it is predicted that elements 58, 15, 14, 47, 35, 26, 68, 59, 36, 25, 24, and 69 have the highest correlation with the specific compression strength and with element 58 ranked the highest and element 69 ranked the lowest for all the subsets. Now, using the first four (58, 15, 14, and 47), five (58, 15, 14, 47, and 35), and seven (58, 15, 14, 47, 35,

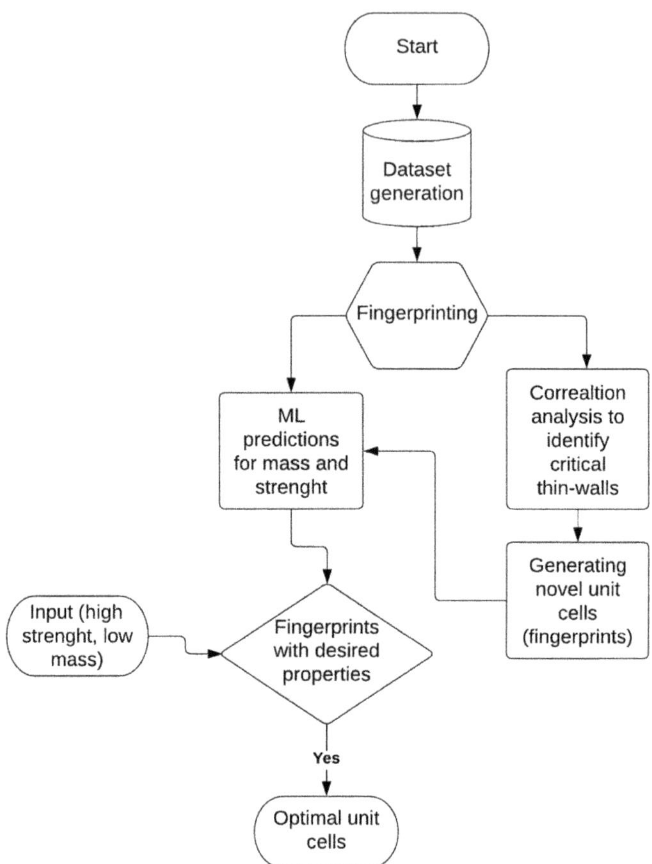

FIGURE 7.12 Flow chart for inverse design. The step-by-step process to reproduce the optimization process includes the following. Step 1: data generation – appropriate training dataset with both input (structures) and output (mechanical properties) should be prepared. Step 2: fingerprinting – the data should be fingerprinted into a machine-readable numerical pattern or sequence. Step 3: ML regression training – once the fingerprinting is done, the training dataset shall be used to train and test suitable ML algorithms that can produce reasonable predictions. Step 4: correlation analysis – to design optimal unit cells, small subsets of the dataset shall be used to find critical walls that have a high influence on the desired properties. These thin walls shall be used to form different unit cells, and the ML algorithm shall be used to predict their mechanical properties. Step 5: optimum unit cell extraction – finally, based on the desired properties, optimal unit cells shall be extracted from the designs developed using correlation analysis and ML predictions.

26, and 68) elements, orthotropic thin-walled unit cells as shown in the following sections are designed and evaluated. For example, the first four elements (58, 15, 14, and 47) lead to the formation of a fingerprint (14 47 12 23 15 58 56) named as "2" in Figure 7.13. Here, elements (12) and (23) are by default considered to form orthotropy with elements (14) and (47). Similarly, element (56) is by default used

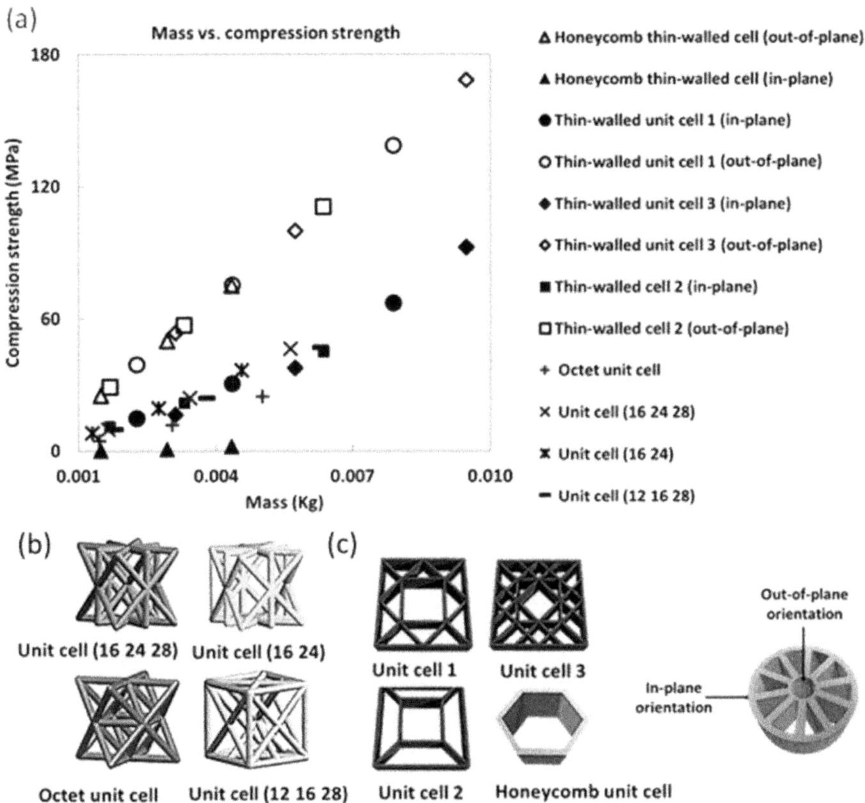

FIGURE 7.13 (a) Numerical comparisons of various lattice and thin-walled cellular structures (in-plane and out-of-plane) under uniaxial compression with an increase in mass as a function of truss diameter or thin-wall thickness using ANSYS simulations, (b) lattice unit cells, and (c) thin-walled cellular unit cells. All the unit cells are designed with the same overall volume and varying rod diameters and wall thicknesses. Panel (a) shows that the compression strength of all the unit cells increases with rod diameter and wall thickness. Especially, while lattice unit cells and thin-walled unit cells in the in-plane orientation have similar strengths, thin-walled unit cells in the out-of-plane orientation exhibit superior strength over a similar mass range.

to form orthotropic with element (58). For the analytical representation of the optimization statement, it can be perceived as follows:

$$\text{yfit} = \text{trainedModel.predictFcn}(\text{Corel1}, \text{Corel2}...\text{Coreln}), \tag{7.4}$$

$$\text{Max}(\text{Corel 1}, \text{Corel 2}...\text{Corel n}) = \text{global optima}, \tag{7.5}$$

where "yfit" is the function used to predict the mechanical properties of new fingerprints generated using correlation analysis and Corel 1, Corel 2 … Corel n were several subsets of fingerprints that were generated through correlation analysis discussed in the previous paragraph. The maximum of each subset generated by the correlation analysis can be perceived as a local optimal solution until no further improvement in the mechanical properties can be achieved, which is where the optimal solution can be perceived as global optima within the dataset or RVE.

To validate the performance of this framework, we manually filtered the entire dataset using the Python command prompt to hard code and extract the optimal thin-walled unit cells. It is observed that the unit cell which is bending-dominated and orthotropic in nature with the highest specific strength within the RVE is the unit cell named "1" in Figure 7.13. This unit cell is among the proposed unit cells using the inverse design framework. Hence, this framework can be considered viable for this type of optimization problem (Table 7.3).

TABLE 7.3

Maxwells Criterion Validation [140, 141]

Lattice Truss Unit Cells		
Octet	$j = 14$ $b = 36$ $M = 36 - 42 + 6 = 0$	Stretch-dominated (M = 0)
16 24	$j = 23$ $b = 48$ $M = 48 - 69 + 6 = -15$	Bending-dominated (M < 0)
16 24 28	$j = 22$ $b = 42$ $M = 42 - 66 + 6 = -12$	Bending-dominated (M < 0)
12 16 28	$j = 23$ $b = 48$ $M = 48 - 69 + 6 = -15$	Bending-dominated (M < 0)
Thin-walled Unit Cells		
1	$j = 24$ $b = 36$ $M = 36 - 48 + 3 = -9$	Bending-dominated (M < 0)
2	$j = 24$ $b = 28$ $M = 28 - 48 + 3 = -17$	Bending-dominated (M < 0)
3	$j = 24$ $b = 36$ $M = 44 - 48 + 3 = -1$	Bending-dominated (M < 0)
Honeycomb	$j = 11$ $b = 12$ $M = 12 - 22 + 3 = -7$	Bending-dominated (M < 0)

7.7 VALIDATION OF CELLULAR UNIT CELLS AND LATTICE UNIT CELLS

To validate the propositions and models, we fabricated several optimal lattice unit cells (Figure 7.14(a)), 4 × 4 lattice structures (Figure 7.14(b)), thin-walled unit cells (Figure 7.15(a)), and 4 × 4 thin-walled cellular structures (Figure 7.15(b)) using AM. Since the goal is to finally propose structures with superior recovery stress based on their strength and flexibility, the lattice and thin-walled unit cells from the previous section are 3D-printed using an SMP. All the unit cells are designed to be of uniform height and varying element diameter and wall thickness. The dimensions of the lattice unit cells are 10 × 10 × 10 mm and the thin-walled unit cells are 10 × 10 × 4 mm. The 4 × 4 lattice structures are 20 × 20 × 20 mm, and the 4 × 4 thin-walled cellular structures are 40 × 40 × 10 mm. To compare the performance of the thin-walled unit cells with the bulk polymer, solid cylinders (diameter 8 mm and height 15 mm) were also 3D-printed. The SMP used in this study is fabricated by combining tris[2-(acryloyloxy) ethyl] isocyanurate (60%) and EPON 826 resin (40%). An open material digital light processing (DLP) AM system (Bison 1000) is used to print all the structures at a printing temperature of 40°C.

An MTS machine (QTEST 150 machine, MTS, USA) with a heating chamber is used to conduct the shape memory programming and stress recovery tests. The chamber is preheated to 75°C (bulk polymer glass transition temperature about 70°C) about 1 hour before the training process to avoid erroneous readings due to the thermal expansions in the fixtures. Once the chamber is heated and ready, the samples are maintained in the chamber for 30 minutes to reach the rubbery state. After that, the samples are compression programmed to 15% strain at a displacement rate of 0.5 mm/min. Once reaching the set strain percentage, the samples are fixed at the compressed shape by rapidly cooling down to room temperature and holding the strain constant. Once at room temperature, the load is removed to fix a temporary shape, and it is observed that the shape fixity ratio (Equation 7.6) for all the structures is almost 100%. Later, the recovery stress for each sample is recorded from the load cell by reheating the samples back to 75°C while maintaining zero recovery strain.

$$F = \frac{\varepsilon_f}{\varepsilon_l} \times 100\%, \tag{7.6}$$

where ε_f is the fixed strain after load removal and ε_l is the measured strain before load removal.

The optimal lightweight cellular unit cells proposed using the inverse design framework in the previous sections along with the lattice unit cells extracted from Refs. [149,150] were modeled and 3D-printed for numerical and experimental validations. Numerical comparisons using the ANSYS simulation tool along with the experimental validations are presented in Figure 7.14 for lattice structures and Figure 7.15 for thin-walled cellular structures. The proposed lattice structures,

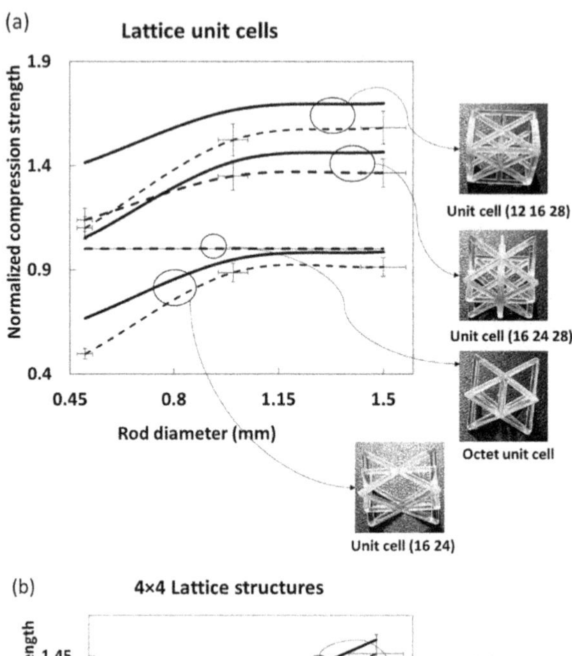

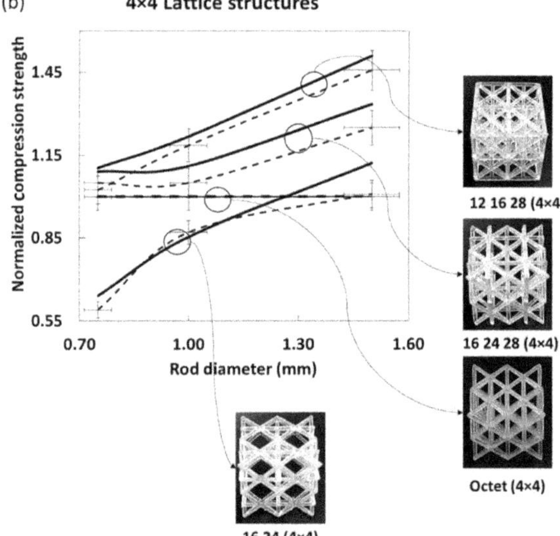

FIGURE 7.14 Experimental (solid lines) and numerical (dashed lines) comparisons for (a) lattice unit cells and (b) 4×4 lattice structures under uniaxial compression test. The thin solid cross represents the error bars from experiments. The thin circle covering the numerical and experimental lines indicates that these two lines belong to the same unit cell, as indicated by the arrow. With the same overall volume, the optimal lattice unit cells (12 16 28) and (16 24 28) can be seen to exhibit 10–60% higher compression strength compared to the octet unit cell in (a). The 4×4 lattice structures in (b) formed by using the unit cells in (a) can also be seen to follow a similar pattern as their unit cells. Here, normalized compressive strength is the ratio of the compressive strength of each lattice unit cell and the compressive strength of the octet unit cell.

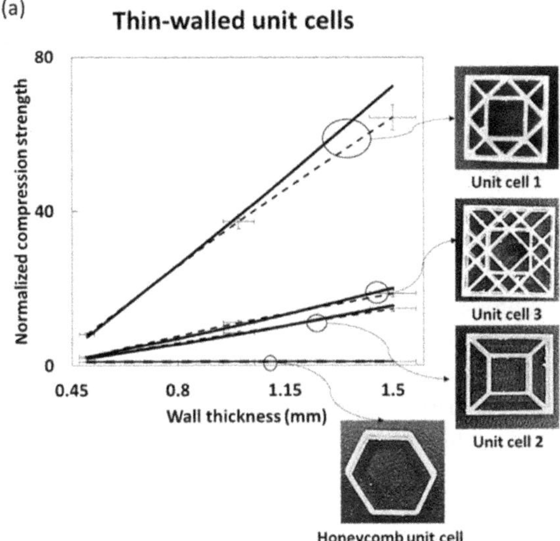

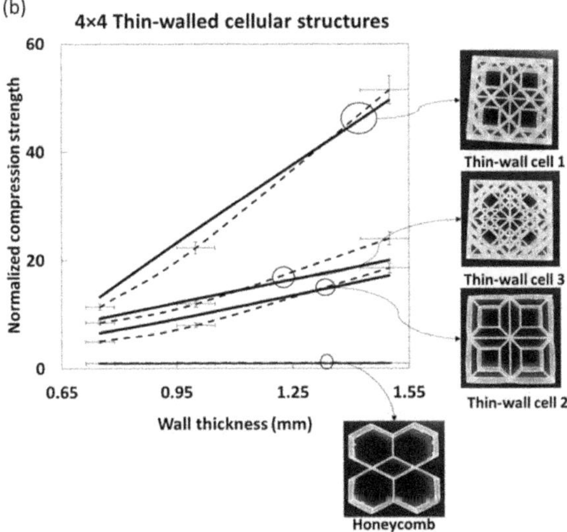

FIGURE 7.15 Experimental (solid lines) and numerical (dashed lines) comparisons for (a) thin-walled unit cells and (b) 4 × 4 thin-walled cellular structures under uniaxial compression tests. The thin solid cross represents the error bars from experiments. The thin circle covering the numerical and experimental lines indicates that these two lines belong to the same unit cell, as indicated by the arrow. The thin-walled unit cells such as unit cell 1, unit cell 2, and unit cell 3 in (a) as well as the 4 × 4 cellular structures formed using the same unit cell can be seen to exhibit much superior compression strength properties compared to honeycomb structure. The superiority of the optimal unit cells can be attributed to their joint connectivity and wall orientations. The normalized compressive strength is the ratio of the compressive strength of individual thin-walled cellular unit cells over that of the honeycomb unit cell.

while are still bending-dominated, can be seen to exhibit similar or even better relative compression strength properties compared to the classical octet truss structure which is stretching-dominated in nature. In Figure 7.14, it should be noted that the comparisons were made with respect to the rod diameters of the lattice structures. Many studies prove that the performance of lattice structures is vastly dependent on their relative densities [139–142]. While the intention of Figure 7.14 is to give a normalized comparison over rod diameters, comparisons for the same optimal lattice structures with octet structures over relative densities were discussed previously in Refs. [149,150]. With respect to relative densities, the optimal lattice unit cells rendered in this study still exhibit superior specific compression strength properties.

The optimal thin-walled cellular unit cells in Figure 7.15(a) can also be seen to be exceptionally superior to honeycomb unit cells in terms of compressive strength (in-plane orientation). These unit cells due to their local and global bending-like behavior will possess flexibility or larger displacements as demonstrated in Figure 7.16. The bending-dominated thin-walled structure can be seen

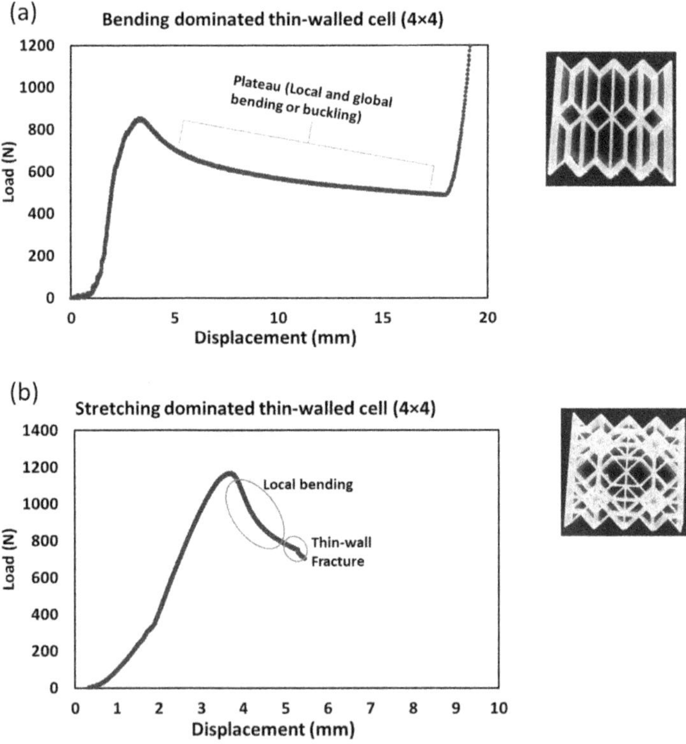

FIGURE 7.16   Test results of bending- and stretch-dominated structures with 4 × 4 unit cells: (a) load-displacement curves of a bending-dominated thin-walled structure and (b) load-displacement curves of a stretch-dominated thin-walled structure (4 × 4 unit cells).

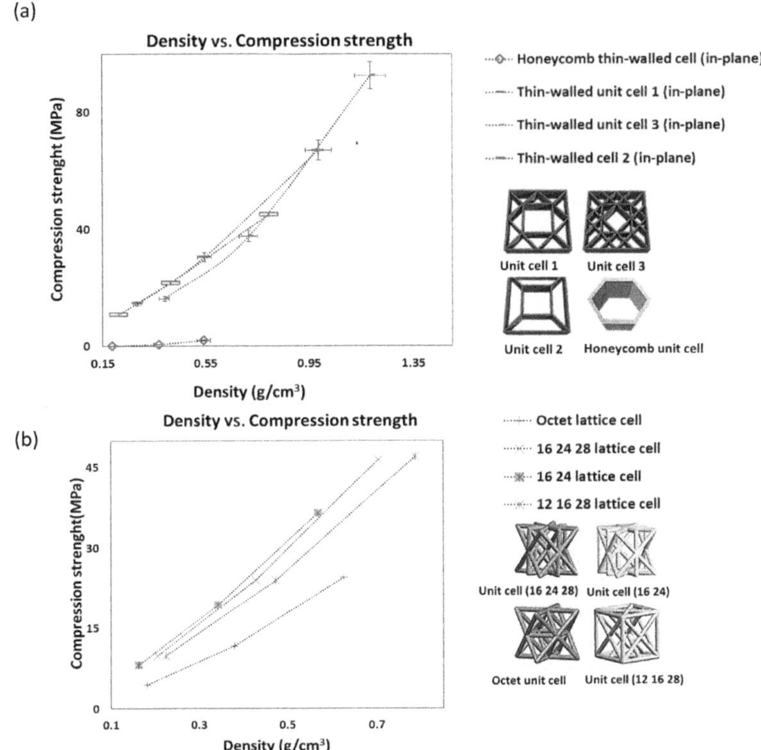

FIGURE 7.17   Change of compressive strength with density for (a) thin-walled cellular unit cells and (b) lattice unit cells.

to exhibit dominant overall bending behavior involving local and global bending, leading to larger displacement and flexibility, while the stretching-dominated structure exhibit global and local buckling and stretching-like behavior with fracture of thin walls.

It can also be observed that the 4 × 4 lattice in Figure 7.14(b) and cellular structures in Figure 7.15(b) exhibit similar properties to that of their unit cells. Figure 7.17 shows the effect of density on the compression strength for thin-walled cellular unit cells and lattice unit cells. Both types of unit cells exhibit an increase in compression strength as the density of the unit cells increases.

7.8   STRESS RECOVERY ANALYSIS FOR DISCOVERED MECHANICAL METAMATERIALS

Using the inverse design framework, we have discovered several new lattice unit cells and thin-walled cellular unit cells and 3D-printed them using an SMP, as discussed earlier. The images and names of the discovered new unit cells are shown in Figure 7.18.

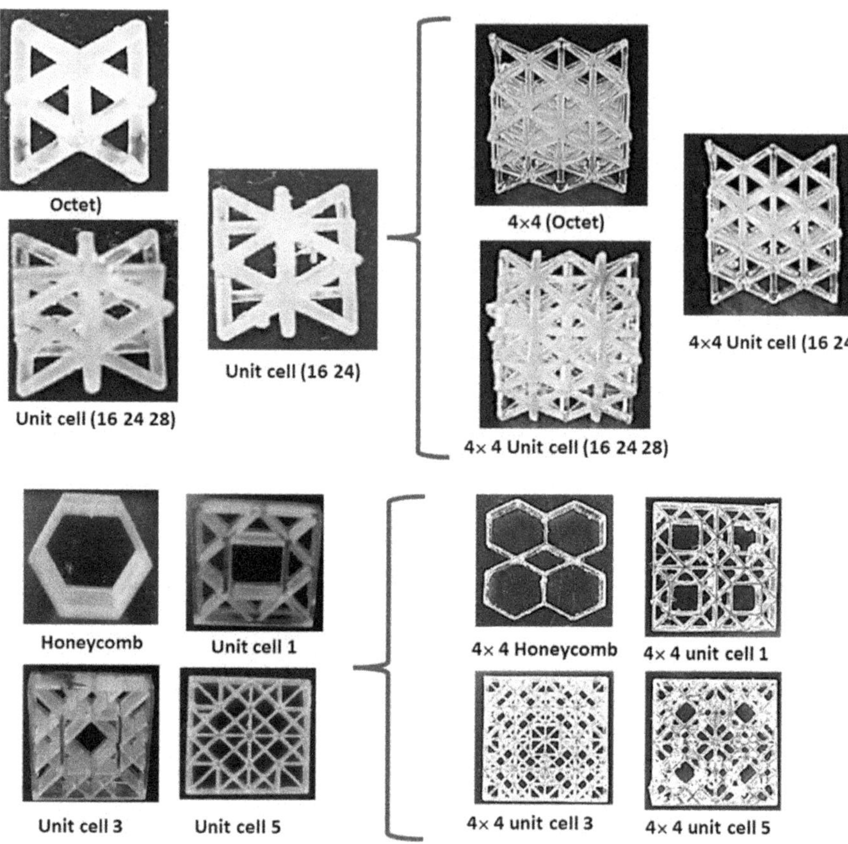

FIGURE 7.18 Newly discovered unit cells (top: lattice unit cells and bottom: thin-walled cellular unit cells).

The comparisons for the optimal lattice unit cells with octet lattice, the proposed thin-walled unit cells with honeycomb unit cell, and the thin-walled unit cell with the solid samples are presented in Figures 7.19 and 7.20, respectively.

As can be seen from Figure 7.19(a) and (b), the optimal bending-dominated lattice unit cells and 4×4 lattice structures, especially unit cell (12 16 28) and unit cell (16 24 28), have about 10–30% higher specific recovery stress (recovery stress/overall volume) compared to octet unit cell with the same lattice member diameter. It should be noted that the Octet lattice unit cell and $4 \times 4 \times 4$ octet lattice structure are stretching-dominated in nature. The optimized lattice unit cells are bending-dominated in behavior. The optimized $4 \times 4 \times 4$ lattice structures, which are either bending-dominated or partially bending-dominated, also exhibit superior recovery stress compared to octet unit cells and $4 \times 4 \times 4$ octet lattice structures.

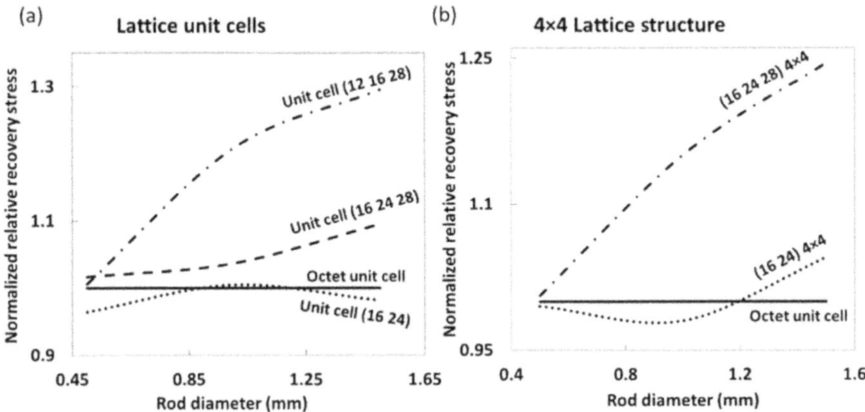

FIGURE 7.19 Experimental comparisons for the normalized recovery stress of (a) lattice unit cells and (b) 4 × 4 lattice structures. Here, the normalized recovery stress is the ratio of the specific recovery stress of each lattice unit cell and the specific recovery stress of the octet unit cell under uniform overall volume. The lattice structures when 3D-printed using an SMP can be seen to exhibit superior (by 30%) stress recovery properties compared to the octet unit cell.

From Figure 7.20(a) and (b), the recovery stress of the optimal thin-walled unit cells and 4 × 4 cellular structures (unit cells 1 and 3) is 200–1000% more than that of the honeycomb unit cell; their recovery stress is also 50% more than that of an optimal stretching-dominated (unit cell 5) structure (see Figure 7.18 for 3D-printed sample images). Figure 7.20(c) shows the comparisons of the mass normalized recovery stress of the optimal thin-walled unit cells and solid material with varying mass, and the optimal unit cells can be seen to exhibit 140–200% higher recovery stress than the solid structure. Similar to lattice unit cells and lattice structures in Figure 7.19, the optimized thin-walled unit cells and 4 × 4 × 4 cellular structures, despite being bending-dominated, still exhibit better recovery stress compared to their stretching-dominated counterparts. This shows that the bending-dominated unit cells can be potential candidates with multifunctional capabilities like superior strength, flexibility, and shape memory.

In Figures 7.19 and 7.20, the effect of rod diameter and wall thickness on the normalized relative recovery stress is presented and discussed. In Figure 7.21, the effect of the normal density of the various unit cells on the normalized relative recovery stress is presented. It is seen that for both types of unit cells, the newly discovered bending-dominated unit cells exhibit higher normalized relative recovery stress than their corresponding nature counterparts, i.e., octet unit cell and honeycomb unit cell.

Also, since the optimization process is based on the structural behavior of the unit cells only, it should be noted that using different SMPs would influence their structural performance and shape memory effect. In this study, the SMP used is brittle in nature (at room temperature) which could lead to less overall

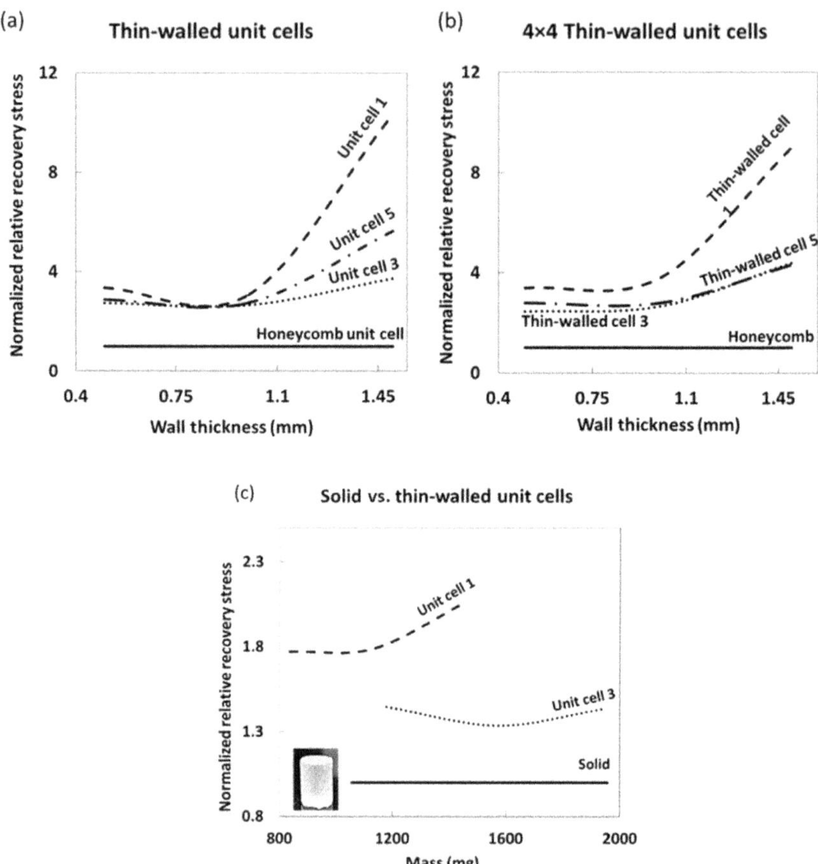

FIGURE 7.20 Experimental comparisons for normalized recovery stress of (a) thin-walled unit cells, (b) 4 × 4 thin-walled cellular structures, and (c) thin-walled unit cell comparison with nonporous solid. Here, the normalized recovery stress is the ratio of the specific recovery stress of each thin-walled structure and the specific recovery stress of the honeycomb structure under uniform overall volume. Here, the honeycomb structure and thin-walled unit cells 1 and 3 were categorized as bending-dominated. The thin-walled unit cell 5 is categorized as stretching-dominated. From (a) and (b), compared to the honeycomb unit cell, the optimized cellular unit cells are exponentially superior in terms of normalized recovery stress. The proposed optimal bending-dominated structures (both the unit cells and the 4 × 4 structures) exhibit similar or even higher recovery stress properties (~50%) compared to the optimal stretching-dominated thin-walled structure.

displacements before fracture. A more ductile SMP can improve the range of displacements when training or programming the structures. A more ductile SMP may also need to consider the nonlinear behavior and viscoelasticity and viscoplasticity during our finite element modeling to create the training dataset. In this study, because all the lattice unit cells and thin-walled cellular unit cells used the

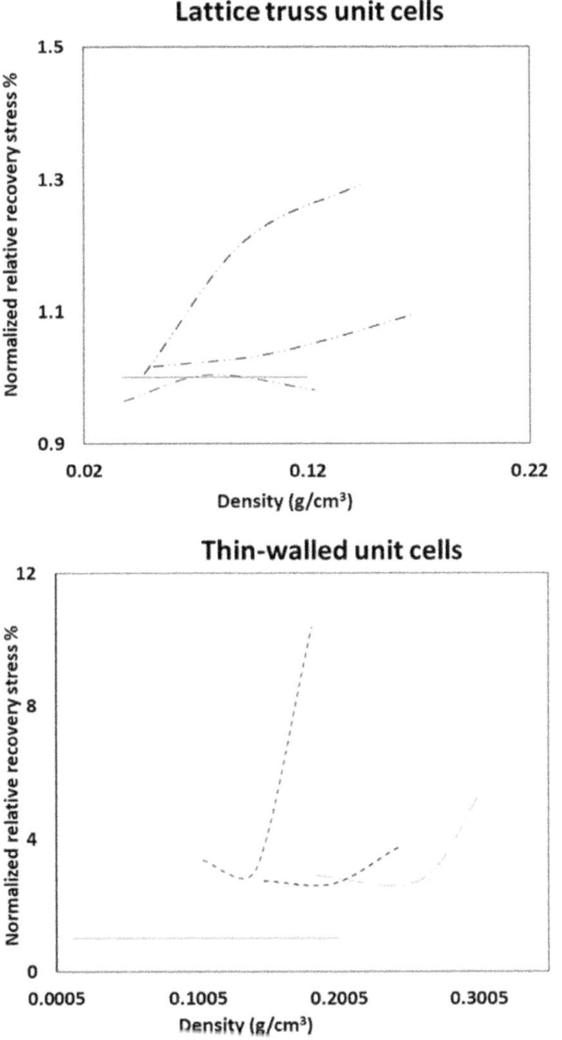

FIGURE 7.21 Normalized relative recovery stress versus density of (a) lattice unit cells and (b) thin-walled unit cells.

same brittle SMP in our modeling and experiments, it is believed that the conclusions may not be changed should another ductile SMP be used.

7.9 SUMMARY AND CONCLUSION

This chapter delves into the intricate realm of 4D printing and its profound implications in engineering, specifically focusing on the burgeoning field of SMPs. Since its inception in 2013, 4D printing has revolutionized traditional 3D printing by introducing the dimension of time, enabling objects to alter their shape or properties in response to external stimuli. This transformative capability, facilitated primarily by SMPs, has sparked significant interest across various engineering sectors.

The paramount importance of recovery stress in 4D-printed structures, particularly in applications necessitating self-actuation and adaptive behavior, drives the study to discover new metamaterials with higher recovery stress. Recovery stress, which denotes a structure's ability to overcome constraints and do positive work to surroundings, plays a critical role in the applications of mechanical metamaterials. Achieving optimal recovery stress entails designing structures that balance flexibility for substantial displacements during programming and strength for storing higher programming stress.

Several studies discussed in the chapter explore diverse techniques to enhance the shape memory properties of SMPs. These techniques range from temperature variations to incorporating nanocomposites, from entropy reduction to enthalpy increase, each aiming to optimize recovery stress while addressing the inherent trade-offs between stiffness and recovery capabilities.

A significant focus of the chapter lies in the exploration of bending-dominated lattice structures as promising candidates for achieving superior recovery stress. These structures exhibit remarkable flexibility, making them adept at accommodating large displacements during programming. Additionally, periodic hexagonal honeycomb structures are investigated for their high shear strength and suitability in applications such as passive morphing airfoils. Furthermore, the chapter introduces a novel optimization framework leveraging machine learning regression models and statistical analysis to discover optimal thin-walled cellular unit cells and lattice unit cells with enhanced recovery stress properties. Experimental validation confirms the efficacy of the proposed framework in identifying structures with superior recovery stress characteristics. It is believed that the inverse design framework proposed in this chapter would have good potential for applications in discovering other types of structures with better mechanical properties and functionalities.

REFERENCES

1. David L, Bourell MC, Leu D, Rosen W. Roadmap for Additive Manufacturing-Identifying the Future of Freeform Processing. Edited and Published by Austin: University of Texas, (2009).
2. Compton BG, Lewis JA. 3D-printing of lightweight cellular composites. Advanced Materials, 26:5930–5935, (2014).
3. Kabir SMF, Mathur K, Seyam AFM. A critical review on 3D printed continuous fiber-reinforced composites: History, mechanism, materials and properties. Composite Structures, 232:111476, (2020).
4. Tibbits S. The emergence of "4D printing". TED Conference, (2013).
5. Tibbits S. 4D printing: Multi-material shape change. Architectural Design, 84:116–121, (2014).

6. Khoo ZX, Teoh JEM, Liu Y, Chua CK, Yang S, An J, Leong KF, Yeong WY. 3D printing of smart materials: A review on recent progresses in 4D printing. Virtual and Physical Prototyping, 10:103–122, (2015).

7. Momeni F, Hassani N, Liu X, Ni J. A review of 4D printing. Materials and Design, 122:42–79, (2017).

8. Melly SK, Liu L, Liu Y, Leng J. On 4D printing as a revolutionary fabrication technique for smart structures. Smart Materials and Structures, 29:083001, (2020).

9. Kuang X, Roach DJ, Wu JT, Hamel CM, Ding Z, Wang TJ, Dunn ML, Qi HJ. Advances in 4D printing: Materials and applications. Advanced Functional Materials, 29:1805290, (2019).

10. Yang C, Boorugu M, Dopp A, Ren J, Martin R, Han D, Choi W, Lee H. 4D printing reconfigurable, deployable and mechanically tunable metamaterials. Materials Horizons, 6:1244–1250, (2019).

11. Kuang X, Chen K, Dunn CK, Wu J, Li VCF, Qi HJ. 3D printing of highly stretchable, shape-memory, and self-healing elastomer toward novel 4D printing. ACS Applied Materials & Interfaces, 10:7381–7388, (2018).

12. Invernizzi M, Turri S, Levi M, Suriano R. 4D printed thermally activated self-healing and shape memory polycaprolactone-based polymers. European Polymer Journal, 101:169–176, (2018).

13. Momeni F, Ni J. Nature-inspired smart solar concentrators by 4D printing. Renewable Energy, 122:35–44, (2018).

14. Zolfagharian A, Kaynak A, Khoo SY, Kouzani A. Pattern-driven 4D printing. Sensors and Actuators A, 274:231–243, (2018).

15. Bodaghi M, Damanpack AR, Liao WH. Triple shape memory polymers by 4D printing. Smart Materials and Structures, 27:065010, (2018).

16. Liu G, Zhao Y, Wu G, Lu J. Origami and 4D printing of elastomer-derived ceramic structures. Science Advances, 4, eaat0641: (2018).

17. Choong YYC, Maleksaeedi S, Eng B, Wei J, Su PC. 4D printing of high performance shape memory polymer using stereolithography. Materials and Design, 126:219–225, (2017).

18. Ding Z, Yuan C, Peng X, Wang T, Qi HJ, Dunn ML. Direct 4D printing via active composite materials. Science Advances, 3:e1602890, (2017).

19. Yang H, Leow WR, Wang T, Wang J, Yu J, He K, Qi D, Wan C, Chen X. 3D printed photoresponsive devices based on shape memory composites. Advanced Materials, 29:1701627, (2017).

20. Zarek M, Mansour N, Shapira S, Cohn D. 4D printing of shape memory-based personalized endoluminal medical devices. Macromolecular Rapid Communications, 38:1600628, (2017).

21. Huang L, Jiang R, Wu J, Song J, Bai H, Li B, Zhao Q, Xie T. Ultrafast digital printing toward 4D shape changing materials. Advanced Materials, 29:1605390, (2017).

22. Zarek M, Layani M, Cooperstein I, Sachyani E, Cohn D, Magdassi S. 3D printing of shape memory polymers for flexible electronic devices. Advanced Materials, 28:4449–4454, (2016).

23. Wu J, Yuan C, Ding Z, Isakov M, Mao Y, Wang T, Dunn ML, Qi HJ. Multi-shape active composites by 3D printing of digital shape memory polymers. Scientific Reports, 6:24224, (2016).

24. Gladman AS, Matsumoto EA, Nuzzo RG, Mahadevan L, Lewis JA. Biomimetic 4D printing. Nature Materials, 15:413–419, (2016).

25. Ge Q, Dunn CK, Qi HJ, Dunn ML. Active origami by 4D printing. Smart Materials and Structures, 23:094007, (2014).

26. Ge Q, Qi HJ, Dunn ML. Active materials by four-dimension printing. Applied Physics Letters, 103:131901, (2013).

27. Li A, Challapalli A, Li G. 4D printing of recyclable lightweight architectures using high recovery stress shape memory polymer. Scientific Reports, 9:7621, (2019).

28. Abedin R, Feng X, Pojman J Jr., Ibekwe A, Mensah P, Warner I, Li G. A thermoset shape memory polymer based syntactic foam with flame retardancy and 3D printability. ACS Applied Polymer Materials, 4:1183–1195, (2022).

29. Ren LQ, Wu Q, Li JY, He YL, Zhang YL, Zhou XL, Wu SY, Liu QP, Li BQ. 4D printing of customizable and reconfigurable mechanical metamaterials. International Journal of Mechanical Sciences, 270:109112, (2024).

30. Lin C, Xin XZ, Tian LF, Zhang D, Liu LW, Liu YJ, Leng JS. Thermal-, magnetic-, and light-responsive 4D printed SMP composites with multiple shape memory effects and their promising applications. Composites Part B: Engineering, 274:111257, (2024).

31. Isaac CW, Duddeck F. Recent progress in 4D printed energy-absorbing metamaterials and structures. Virtual and Physical Prototyping, 18:e2197436, (2023).

32. Wu Y, Han Y, Wei ZX, Xie Y, Yin J, Qian J. 4D printing of chiral mechanical metamaterials with modular programmability using shape memory polymer. Advanced Functional Materials, 33:2306442, (2023).

33. Owens CAH, Wang YP, Farzinazar S, Yang C, Lee HW, Lee JH. Tunable thermal transport in 4D printed mechanical metamaterials. Materials & Design, 231:111992, (2023).

34. Ma LH, Wei TY, Rao W, Zhang K, Gao H, Chen XJ, Zhang XC. 4D printed chiral metamaterials with negative swelling behavior. Smart Materials and Structures, 32:015014, (2023).

35. Peng XR, Wu S, Sun XH, Yue L, Montgomery SM, Demoly F, Zhou K, Zhao RR, Qi HJ. 4D printing of freestanding liquid crystal elastomers via hybrid additive manufacturing. Advanced Materials, 34:2204890, (2022).

36. Bodaghi M, Damanpack AR, Liao WH. Adaptive metamaterials by functionally graded 4D printing. Materials & Design, 135:26–36, (2017).

37. Xin XZ, Liu LW, Liu YJ, Leng JS. 4D printing auxetic metamaterials with tunable, programmable, and reconfigurable mechanical properties. Advanced Functional Materials, 30:2004226, (2020).

38. Yan C, Li G. Design oriented constitutive modeling of amorphous shape memory polymers and its application to multiple length scale lattice structures. Smart Materials and Structures, 28:095030, (2019).

39. Tao R, Ji LT, Li Y, Wan ZS, Hu WX, Wu WW, Liao BB, Ma LH, Fang DN. 4D printed origami metamaterials with tunable compression twist behavior and stress-strain curves. Composites Part B: Engineering, 201:108344, (2020).

40. Huang S, Chen Y. 4D printing: A new trend in three-dimensional printing. Assemble, 51:12–14, (2013).

41. Wu J, Yuan C, Peng Y, Guo Z, Zhang H. 4D printing of shape memory polymers with multiple programmable actuators. Journal of Materials Chemistry B, 7:2885–2891, (2019).

42. Ge Q, Qi HJ, Dunn ML. Volumetric additive manufacturing via tomographic reconstruction. Science Advances, 2:e1600324, (2016).

43. Wallin TJ, Pikul J, Shepherd RF. 3D printing of soft robotic systems. Nature Review Materials, 3:84–100, (2018).

44. Saadi MASR, Maguire A, Pottackal NT, Thakur MSH, Ikram MM, Hart AJ, Ajayan PM, Rahman MM. Direct ink writing: A 3D printing technology for diverse materials. Advanced Materials, 34:2108855, (2022).

45. Mu QY, Wang L, Dunn CK, Kuang X, Duan F, Zhang Z, Qi HJ, Wang TJ. Digital light processing 3D printing of conductive complex structures. Additive Manufacturing, 18:74–83, (2017).

46. Zhang B, Li HG, Cheng JX, Ye HT, Sakhaei AH, Yuan C, Rao P, Zhang YF, Chen Z, Wang R, He XN, Liu J, Xiao R, Qu SX, Ge Q. Mechanically robust and UV-curable shape-memory polymers for digital light processing based 4D printing. Advanced Materials, 33:2101298, (2021).

47. Caprioli M, Roppolo I, Chiappone A, Larush L, Pirri CF, Magdassi S. 3D-printed self-healing hydrogels via digital light processing. Nature Communications, 12:2462, (2021).

48. Zhang Y, Huang LM, Song HJ, Ni CJ, Wu JJ, Zhao Q, Xie T. 4D printing of a digital shape memory polymer with tunable high performance. ACS Applied Materials & Interfaces, 11:32408–32413, (2019).

49. Rossegger E, Höller R, Reisinger D, Strasser J, Fleisch M, Griesser T, Schlögl S. Digital light processing 3D printing with thiol-acrylate vitrimers. Polymer Chemistry, 12:639–644, (2021).

50. Li HG, Zhang BA, Wang R, Yang XD, He XN, Ye HT, Cheng JX, Yuan C, Zhang YF, Ge Q. Solvent-free upcycling vitrimers through digital light processing-based 3D printing and bond exchange reaction. Advanced Functional Materials, 32:2111030, (2022).

51. Madrid-Wolff J, Toombs J, Rizzo R, Bernal PN, Porcincula D, Walton R, Wang B, Kotz-Helmer F, Yang Y, Kaplan D, Zhang YS, Zenobi-Wong M, Mcleod RR, Rapp B, Schwartz J, Shusteff M, Talyor H, Levato R, Moser C. A review of materials used in tomographic volumetric additive manufacturing. MRS 50th Anniversary Prospective, 13:764–785, (2023).

52. Wang X, Xu S, Zhou S, Xu W, Leary M, Choong P, Qian M, Brandt M, Xie YM. Topological design and additive manufacturing of porous metals for bone scaffolds and orthopaedic implants: A review. Biomaterials, 83:127–141, (2016).

53. Bhargav A, Sanjairaj V, Rosa V, Feng LW, Fuh YH. Applications of additive manufacturing in dentistry: A review. Journal of Biomedical Materials Research Part B: Applied Biomaterials, 106:2058–2064, (2018).

54. Nguyen DT, Meyers C, Yee TD, Dudukovic NA, Destino JF, Zhu C, Duoss EB, Baumann TF, Suratwala T, Smay JE, Dylla-Spears R. 3D-printed transparent glass. Advanced Materials, 29:1701181, (2017).

55. Gong H, Bickham BP, Woolley AT, Nordin GP. Custom 3D printer and resin for 18 μm × 20 μm microfluidic flow channels. Lab on Chip, 17:2899–2909, (2017).

56. Ngo TD, Kashani A, Imbalzano G, Nguyen KTQ, Hui D. Additive manufacturing (3D printing): A review of materials, methods, applications and challenges. Composites Part B: Engineering, 143:172–196, (2018).

57. Kelly BE, Bhattacharya I, Heidari H, Shusteff M, Spadaccini SM, Taylor HK. Volumetric additive manufacturing via tomographic reconstruction. Science, 363:1075–1079, (2019).

58. Sachs E, Wylonis E, Allen S, Cima M, Guo H. Production of injection molding tooling with conformal cooling channels using the three dimensional printing process. Polymer Engineering and Science, 40:1232–1247, (2000).

59. Shusteff M, Browar AEM, Kelly BE, Henriksson J, Weisgraber TH, Panas RM, Fang NX, Spadaccini CM. One-step volumetric additive manufacturing of complex polymer structures. Science Advance, 3:eaao5496, (2017).

60. Bernal PN, Delrot P, Loterie D, Li Y, Malda J, Moser C, Levato R. Volumetric bioprinting of complex living-tissue constructs within seconds. Advanced Materials, 31:1904209, (2019).

61. Toombs JT, Luitz M, Cook CC, Jenne S, Li CC, Rapp BE, Kotz-Helmer F, Taylor HK. Volumetric additive manufacturing of silica glass with microscale computed axial lithography. Science, 376:308, (2022).
62. Weisgraber TH, de Beer MP, Huang S, Karnes JJ, Cook CC, Shusteff M. Virtual volumetric additive manufacturing (VirtualVAM). Advanced Materials Technology, 8:2301054, (2023).
63. Ahn D, Stevens LM, Zhou K, Page ZA. Rapid high-Resolution visible light 3D printing. ACS Central Science, 6:1555–1563, (2020).
64. Darkes-Burkey C, Shepherd RF. High-resolution 3D printing in seconds. Nature, 588:594–595, (2020).
65. Regehly M, Garmshausen Y, Reuter M, König NF, Israel E, Kelly DP, Chou CY, Koch K, Asfari B, Hecht S. Xolography for linear volumetric 3D printing. Nature, 588:620–627, (2020).
66. Shusteff M, Browar AEM, Kelly BE, Henriksson J, Weisgraber TH, Panas RM, Fang NX, Spadaccini CM. One-step volumetric additive manufacturing of complex polymer structures. Science Advances, 3:eaao5496, (2017).
67. Hart AJ, Rao A. How to print a 3D object all at once: A tomography-based method prints polymer objects volumetrically. Science, 363:1042–1043, (2019).
68. Liu YJ, Du HY, Liu LW, Leng JS. Shape memory polymers and their composites in aerospace applications: A review. Smart Materials and Structures, 23:023001, (2014).
69. Lan X, Liu YJ, Lv HB, Wang XH, Leng JS, Du SY. Fiber reinforced shape-memory polymer composite and its application in a deployable hinge. Smart Materials and Structures, 18:024002, (2009).
70. Mao YQ, Yu K, Isakov MS, Wu JT, Dunn ML, Qi HJ. Sequential self-folding structures by 3D printed digital shape memory polymers. Scientific Reports, 5:13616, (2015).
71. Liu TW, Bai JB, Fantuzzi N, Zhang X. Thin-walled deployable composite structures: A review. Progress in Aerospace Sciences, 146:100985, (2024).
72. Liu W, Kong DY, Zhao W, Leng JS. Multi-stimulus responsive shape memory polyurea incorporating stress-mismatching structure for soft actuators and reversible deployable structures. Composite Structures, 334:117966, (2024).
73. Kang D, Jeong JM, Jeong KI, Kim SS. Improving the deformability and recovery moment of shape memory polymer composites for bending actuators: Multiple neutral axis skins and deployable core. ACS Applied Materials & Interfaces, 15:33944–33956, (2023).
74. Lendlein A, Langer R. Biodegradable, elastic shape-memory polymers for potential biomedical applications. Science, 296:1673–1676, (2002).
75. Lendlein A, Kelch S. Shape-memory polymers as stimuli-sensitive implant materials. Clinical Hemorheology and Microcirculation, 32:105–116, (2005).
76. Li YR, Meng Q, Chen SJ, Ling PX, Kuss MA, Duan B, Wu SH. Advances, challenges, and prospects for surgical suture materials. Acta Biomaterialia, 168:78–112, (2023).
77. Sun SQ, Chen CX, Zhang JH, Hu JS. Biodegradable smart materials with self-healing and shape memory function for wound healing. RSC Advances, 13:3155–3163, (2023).
78. Yakacki CM, Shandas R, Lanning C, Rech B, Eckstein A, Gall K. Unconstrained recovery characterization of shape-memory polymer networks for cardiovascular applications. Biomaterials, 28:2255–2263, (2007).
79. Wache HM, Tartakowska DJ, Hentrich A, Wagner MH. Development of a polymer stent with shape memory effect as a drug delivery system. Journal of Materials Science-Materials in Medicine, 14:109–112, (2003).

80. Van Daele L, Chausse V, Parmentier L, Brancart J, Pegueroles M, Van Vlierberghe S, Dubruel P. 3D-printed shape memory poly(alkylene terephthalate) scaffolds as cardiovascular stents revealing enhanced endothelialization. Advanced Healthcare Materials, (2024). DOI: 10.1002/adhm.202303498.

81. Zhu Y, Deng KC, Zhou JW, Lai C, Ma ZW, Zhang H, Pan JZ, Shen LY, Bucknor MD, Ozhinsky E, Kim S, Chen GJ, Ye SH, Zhang Y, Liu DH, Gao CY, Xu YH, Wang HA, Wagner WR. Shape-recovery of implanted shape-memory devices remotely triggered via image-guided ultrasound heating. Nature Communications, 15:1123, (2024).

82. Sarrafan S, Li G. Conductive and ferromagnetic syntactic foam with shape memory and self-Healing/Recycling capabilities. Advanced Functional Materials, 34:2308085, (2024).

83. Mahmud S, Konlan J, Deicaza J, Li G. Coir/glass hybrid fiber reinforced thermoset composite laminates with room-temperature self-healing and shape memory properties. Industrial Crops and Products, 201:116895, (2023).

84. Tetteh O, Mensah P, Li G. Repeated healing of low velocity impact induced damage in orthogrid-stiffened Sandwich panel. Journal of Composite Materials, 57:3619–3632, (2023).

85. Konlan J, Feng X, Li G. A multifunctional hybrid extrinsic-intrinsic self-healing laminated composites. Smart Materials and Structures, 32:075006, (2023).

86. Konlan J, Mensah P, Ibekwe S, Li G. A laminated vitrimer composite with strain sensing, delamination self-healing, deicing, and room-temperature shape restoration properties. Journal of Composite Materials, 56:2267–2278, (2022).

87. Feng X, Li G. Room-temperature self-healable and mechanically robust thermoset polymer for healing delamination and recycling carbon fiber. ACS Applied Materials and Interfaces, 13:53099–53110, (2021).

88. Afful HQ, Ibekwe S, Mensah P, Li G. Influence of uniaxial compression on the shape memory behavior of vitrimer composite embedded with tension-programmed unidirectional shape memory polymer fibers. Journal of Applied Polymer Science, 138:50429, (2021).

89. Feng X, Li G. Versatile phosphate diester-based flame retardant vitrimers via catalyst-free mixed transesterification. ACS Applied Materials & Interfaces, 12:57486–57496, (2020).

90. Feng X, Li G. Catalyst-free β-hydroxy phosphate ester exchange for robust fireproof vitrimers. Chemical Engineering Journal, 417:129132, (2021).

91. Feng X, Fan J, Li A, Li G. Biobased tannic acid crosslinked epoxy thermosets with hierarchical molecular structure and tunable properties: Damping, shape memory and recyclability. ACS Sustainable Chemistry & Engineering, 8:874–883, (2020).

92. Feng X, Fan J, Li A, Li G. Multi-reusable thermoset with anomalous flame triggered shape memory effect. ACS Applied Materials & Interfaces, 11:16075–16086, (2019).

93. Lu L, Pan J, Li G. Recyclable high performance epoxy based on transesterification reaction. Journal of Materials Chemistry A, 5:21505–21513, (2017).

94. Lu L, Fan J, Li G. Intrinsic healable and recyclable thermoset epoxy based on shape memory effect and transesterification reaction. Polymer, 105:10–18, (2016).

95. Zhang P, Ogunmekan B, Ibekwe S, Jerro D, Pang SS, Li G. Healing of shape memory polyurethane fiber reinforced syntactic foam subjected to tensile stress. Journal of Intelligent Material Systems and Structures, 27:1792–1801, (2016).

96. Champagne J, Pang SS, Li G. Effects of partial confinement and local heating on healing efficiencies of self-healing particulate composites. Composites Part B: Engineering, 97:344–352, (2016).

97. Shojaei A, Sharafi S, Li G. A multiscale theory of self-crack-healing with solid healing agent assisted by shape memory effect. Mechanics of Materials, 81:25–40, (2015).

98. Li G, Zhang P. A self-healing particulate composite reinforced with strain hardened short shape memory polymer fibers. Polymer, 54:5075–5086, (2013).

99. Li G, Ajisafe O, Meng H. Effect of strain hardening of shape memory polymer fibers on healing efficiency of thermosetting polymer composites. Polymer, 54:920–928, (2013).

100. Li G, Meng H, Hu J. Healable thermoset polymer composite embedded with stimuli-responsive fibers. Journal of the Royal Society Interfaces, 9:3279–3287, (2012).

101. Nji J, Li G. Damage healing ability of a shape memory polymer based particulate composite with small thermoplastic contents. Smart Materials and Structures, 21:025011, (2012).

102. Nji J, Li G. A biomimic shape memory polymer based self-healing particulate composite. Polymer, 51:6021–6029, (2010).

103. Li G, Uppu N. Shape memory polymer based self-healing syntactic foam: 3-d confined thermomechanical characterization. Composites Science and Technology, 70:1419–1427, (2010).

104. John M, Li G. Self-healing of Sandwich structures with grid stiffened shape memory polymer syntactic foam core. Smart Materials and Structures, 19:075013, (2010).

105. Nji J, Li G. A self-healing 3D woven fabric reinforced shape memory polymer composite for impact mitigation. Smart Materials and Structures, 19:035007, (2010).

106. Li G, Nettles D. Thermomechanical characterization of a shape memory polymer based self-repairing syntactic foam. Polymer, 51:755–762, (2010).

107. Li G, John M. A self-healing smart syntactic foam under multiple impacts. Composites Science and Technology, 68:3337–3343, (2008).

108. Yang Q, Li G. A spider silk like shape memory polymer fiber for vibration damping. Smart Materials and Structures, 23:105032, (2014).

109. Sharafi S, Li G. A mutliscale approach for modeling actuation response of polymeric artificial muscle. Soft Matter, 11:3833–3843, (2015).

110. Sharafi S, Li G. Multiscale modeling of vibration damping response of shape memory polymer fibers. Composites Part B-Engineering, 91:306–314, (2016).

111. Zhang P, Ayaugbokor U, Ibekwe S, Jerro D, Pang SS, Mensah P, Li G. Healing of polymeric artificial muscle reinforced ionomer composite by resistive heating. Journal of Applied Polymer Science, 133:43660, (2016).

112. Yang Q, Li G. A top-down multi-scale modeling for actuation response of polymeric artificial muscles. Journal of the Mechanics and Physics of Solids, 92:237–259, (2016).

113. Yang Q, Fan J, Li G. Artificial muscles made of chiral two-way shape memory polymer fibers. Applied Physics Letters, 109:183701, (2016).

114. Zhang P, Li G. Fishing line artificial muscle reinforced composite for impact mitigation and on-demand damage healing. Journal of Composite Materials, 50:4235–4249, (2016).

115. Fan J, Li G. High performance and tunable artificial muscle based on two-way shape memory polymer. RSC Advances, 7:1127–1136, (2017).

116. Fan J, Li G. High enthalpy storage thermoset network with giant stress and energy output in rubbery state. Nature Communications, 9:642, (2018).

117. Li A, Fan J, Li G. Recyclable thermoset shape memory polymer with high stress and energy output via facile UV-curing. Journal of Materials Chemistry A, 6:11479–11487, (2018).

118. Feng X, Li G. high-temperature shape memory photopolymer with intrinsic flame retardancy and record-high recovery stress. Applied Materials Today, 23: 101056, (2021).

119. Wick C, Peters A, Li G. Quantifying the contributions of energy storage in A thermoset shape memory polymer with high stress recovery: A molecular dynamics study. Polymer, 213:123319, (2021).

120. Yan C, Li G. A mechanism based four-chain constitutive model for enthalpy driven thermoset shape memory polymers with finite deformation. Journal of Applied Mechanics-Transactions of ASME, 87:061007, (2020).

121. Li G, Wang A. Cold, warm, and hot programming of shape memory polymers. Journal of Polymer Science Part B: Polymer Physics, 54:1319–1339, (2016).

122. Li G, Shojaei A. A viscoplastic theory of shape memory polymer fibers with application to self-healing materials. Proceedings of the Royal Society A-Mathematical Physical and Engineering Sciences, 468:2319–2346, (2012).

123. Xu W, Li G. Thermoviscoplastic modeling and testing of shape memory polymer based self-healing syntactic foam programmed at glassy temperature. ASME Journal of Applied Mechanics, 78:061017, (2011).

124. Li G, Xu W. Thermomechanical behavior of thermoset shape memory polymer programmed by cold-compression: Testing and constitutive modeling. Journal of the Mechanics and Physics of Solids, 59:1231–1250, (2011).

125. Li C, Wu Y, Guo Y. Enhancement of shape memory effect and recovery stress of shape memory polymer nanocomposites. Journal of Materials Science, 53:5874–5885, (2018).

126. Chen J, Yan D, Zhou Y. Shape memory polyurethane nanocomposites with enhanced recovery stress via synergistic effect of graphene and polyhedral oligomeric silsesquioxane. Journal of Materials Chemistry C, 4:348–357, (2016).

127. Yang Y, Wang Y, Qiu J. Experimental study on the recovery stress of lattice structures made from shape memory polymers. Journal of Materials Science, 55:9834–9844, (2020).

128. Fang X, Wen J, Cheng L, Yu D, Zhang H, Gumbsch P. Programmable gear-based mechanical metamaterials. Nature Materials, 21:869–876, (2022).

129. Zhao Y, Zhang H, Leng J. 3D printed auxetic structures with tunable mechanical properties for medical devices. Materials & Design, 203:109637, (2021).

130. Wang X, Yang Y, Zhang Q. Hierarchically structured metamaterials with strain-dependent solid-solid phase change for micro-actuators and grippers. Acta Mechanica Solida Sinica, 33:228–240, (2020).

131. Yang Y, Zhang H, Liu Y. Review on two-dimensional and three-dimensional auxetic structures. Journal of Materials Science, 54:13047–13068, (2019).

132. Attallah MM, El-Safty S, Mahmoud AM. The unique behavior of auxetic structures and their potential applications. Journal of Materials Science, 55:13699–13718, (2020).

133. Deng K, Zhang Y, Kang Z. Topology optimization of cellular auxetic metamaterials. Journal of Applied Mechanics, 87, 101006: (2020).

134. Evans KE, Nkansah MA, Hutchinson IJ, Rogers SC. Molecular network design. Nature, 404:850–853, (2000).

135. Lakes RS. Foam structures with a negative Poisson's ratio. Science, 235:1038–1040, (1987).

136. Wang Y, Liu Y, Wang X, Zhang P, Cheng H. Shape-memory polymers with high shape recovery strain and its application in a passive morphing airfoil. Smart Materials and Structures, 24:125023, (2015).

137. Meng H, Li G. A review of stimuli-responsive shape memory polymer composites. Polymer, 54:2199–2221, (2013).

138. Yang Q, Li G. Temperature, and rate dependent thermomechanical modeling of shape memory polymers with physics-based phase evolution law. International Journal of Plasticity, 80:168–186, (2016).

139. Evans AG, Hutchinson JW, Ashby MF. Multifunctionality of cellular metal systems. Progress in Materials Science, 43:171–221, (1998).

140. Maxwell JC. On the calculation of the equilibrium and stiffness of frames. Philosophical Magazine, 27:294, (1864).

141. Deshpande VS, Fleck NA, Ashby MF. Foam topology bending vs stretching dominated architecture. Acta Materialia, 49:1035–1040, (2001).

142. Deshpande VS, Fleck NA, Ashby MF. Effective properties of the octet-truss lattice material. Journal of the Mechanics and Physics of Solids, 49:1747–1769, (2001).

143. Zhang P, Heyne MA, To AC. Biomimetic staggered composites with highly enhanced energy dissipation: Modeling, 3D printing, and testing. Journal of the Mechanics and Physics of Solids, 83:285–300, (2015).

144. Panda B, Leite M, Biswal BB, Niu X, Garg A. Experimental and numerical modelling of mechanical properties of 3D printed honeycomb structures. Measurement, 116:495–506, (2018).

145. Sha Y, Jiani L, Haoyu C, Robert OR, Jun X. Design and strengthening mechanisms in hierarchical architected materials processed using additive manufacturing. International Journal of Mechanical Sciences, 149:150–163, (2018).

146. Tan XP, Tan YJ, Chow CSL, Tor SB, Yeong WY. Metallic powder-bed based 3D printing of cellular scaffolds for orthopedic implants: A state-of-the-art review on manufacturing, topological design, mechanical properties, and biocompatibility. Science and Engineering C, 76:1328–1343, (2017).

147. Challapalli A, Li G. Machine learning assisted design of new lattice core for Sandwich structures with superior load carrying capacity. Scientific Reports, 11:18552, (2021).

148. Challapalli A, Li G. 3D printable biomimetic rod with superior buckling resistance designed by machine learning. Scientific Reports, 10:20716, (2020).

149. Challapalli A, Patel D, Li G. Inverse machine learning framework for optimizing lightweight metamaterials. Materials & Design, 208:109937, (2021).

150. Challapalli A, Konlan J, Patel D, Li G. Discovery of cellular unit cells with high natural frequency and energy absorption capabilities by an inverse machine learning framework. Frontiers in Mechanical Engineering, 7:779098, (2021).

151. Agresti A. Categorical Data Analysis. John Wiley & Sons, (2018).

152. Gang Z, Wenlong D, Jing T, Jingshou L, Yang G, Siyu S, Ruyue W, Ning S. Spearman rank correlations analysis of the elemental, mineral concentrations, and mechanical parameters of the lower Cambrian Niutitang Shale: A case study. Journal of Petroleum Science and Engineering (Part, C) 208, 109550: (2022).

153. Lin WT, Wu YC, Cheng A, Chao SJ, Hsu HM. Engineering properties and correlation analysis of fiber cementitious materials. Materials, 7:7423–7435, (2014).

154. Tummala R. Predictive Modeling of FMOL Health System Utilization Using Machine Learning Algorithms and Retrospective Study of COVID Tested Patients, LSU Master's Theses. (2021).

155. Chengwei X, Jiaqi Y, Rui ME, Chunming R. Using Spearman's correlation coefficients for exploratory data analysis on big dataset. Concurrency and Computation: Practice and Experience, 28:3866–3878, (2016).

156. Bonett DG, Wright TA. Sample size requirements for estimating Pearson, Kendall and Spearman correlations. Psychometrika, 65:23–28, (2000).

8 Summary and Future Perspectives

8.1 SUMMARY OF MACHINE LEARNING–ASSISTED DISCOVERY OF MECHANICAL METAMATERIALS

The design of complex lightweight structures or mechanical metamaterials represents a formidable challenge, demanding meticulous consideration of a myriad of factors encompassing strength, rigidity, weight, and cost. In recent years, the integration of machine learning and data analysis has emerged as a pivotal approach in optimizing the design of such structures, ushering in a new era of innovation and efficiency. This book serves as a compendium of pioneering methodologies and novel techniques that harness the power of machine learning to achieve topology optimization, thereby generating lightweight structures that excel in strength, efficiency, and functionality.

From a larger picture of lifecycle management of load-bearing structures, machine learning–assisted discovery of metamaterials is an integral component of this ecosystem. It starts from engineering or design specifications. Based on the specifications, researchers or engineers need to select the proper materials. If such materials cannot be found, then the design of new materials needs to be conducted, for example, discovering new 3D printable shape memory polymers guided by machine learning. After that, the proper structures that meet the engineering specifications need to be designed and optimized, which is the focus of this book. Manufacturing of the designed structures such as additive manufacturing is the next step, with experimental validation. The structure is ready for service, which needs structural health monitoring. If damage occurs during service, repair or damage self-healing needs to be conducted to extend the service life. Finally, at the end of service, the materials used need to be recycled and reused to make the material sustainable. Each step needs to communicate with the central hub through the internet of people (IoP), the internet of things (IoT), and the internet of manufacturing (IoM). Using a lattice-cored polymer composite sandwich structure as an example, the structure lifecycle management is shown in Figure 8.1.

From Figure 8.1, structural design using machine learning is one integral component of the structure lifecycle management. Machine learning algorithms stand at the forefront of this endeavor, tasked with modeling the intricate behavior of structures, predicting their strength and stiffness, and iteratively refining the design process. Leveraging datasets teeming with structures endowed with known strength and stiffness properties, these algorithms undergo rigorous training to hone their predictive capabilities. Through a cyclical process of prediction and optimization, machine learning models prognosticate the strength and stiffness of

DOI: 10.1201/9781003400165-8

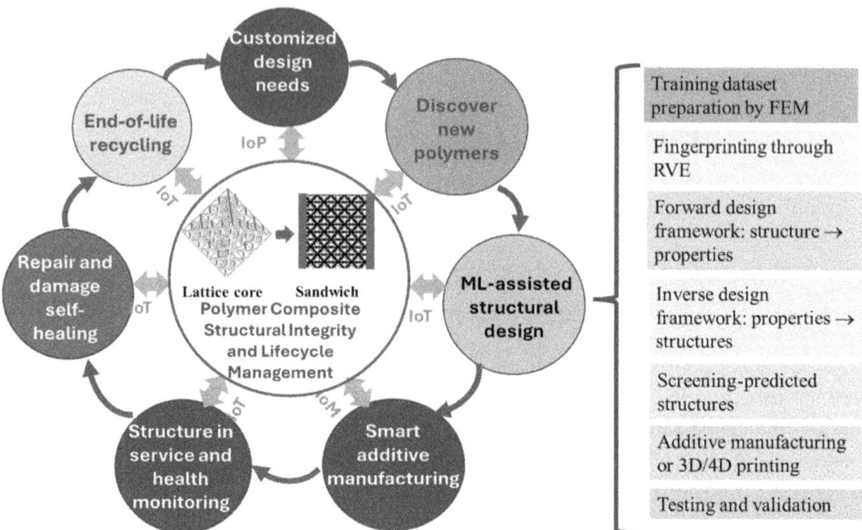

FIGURE 8.1 Schematic of structure lifecycle management and the components involved in machine learning–assisted structural design and optimization.

structures, facilitating subsequent topology optimizations to iteratively enhance the design until the desired specifications are met. This synergistic fusion of machine learning and topology optimization represents a promising avenue for optimizing the design of lightweight structures, offering unprecedented levels of efficiency and efficacy. A general machine learning framework is schematically shown in Figure 8.2.

A diverse array of previously unexplored symmetric and asymmetric isotropic and anisotropic lattice unit cells takes center stage, their designs meticulously crafted and scrutinized through the lens of machine learning. Symmetric optimal lattice unit cells, crafted through this innovative approach, boast compression strengths surpassing those of traditional octet lattice unit cells by substantial margins ranging from 28 to 67%, alongside flexural strengths eclipsing them by 13 to 35%.

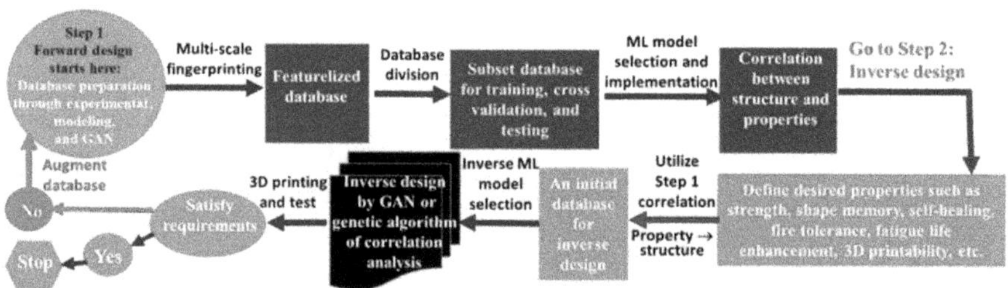

FIGURE 8.2 Schematic of a general machine learning framework.

In a bid to further enhance the performance of these lattice unit cells, biomimetic rods are seamlessly integrated, augmenting their buckling capacity and culminating in remarkable enhancements ranging from 130 to 160% in relative buckling capacities compared to unit cells with solid rods.

Despite the strides made, the proposed methodology remains ripe for further refinement and enhancement. Suggestions for improvement include enlarging the training dataset to bolster prediction accuracy and employing more sophisticated topology optimization algorithms capable of unearthing increasingly optimal designs. The utilization of generative adversarial networks (GANs) and forward regression techniques has paved the way for the proposition of novel lattice unit cells and cellular unit cells that outperform traditional designs, a testament to the transformative potential of machine learning in structural optimization.

The inverse machine learning framework emerges as a beacon of innovation, enabling the continuous optimization of lattice unit cells through iterative refinement fueled by the utilization of newly created unit cells as training data for subsequent predictions. This versatile framework holds promise in optimizing a myriad of structural designs and conceiving novel structures tailored to meet a diverse array of mechanical properties. From shock absorption to enhanced bending or buckling strengths and shape recovery, the applications of this technique are as varied as they are profound, promising to revolutionize the landscape of structural design.

The book also displays the remarkable efficacy of the inverse machine learning technique in designing optimal cellular unit cells that outshine their biomimetic counterparts. Exploration of an expansive design space, yielding nearly 750,000 designs, has led to the proposition of novel designs boasting superior structural properties across a multitude of metrics. These optimized cellular unit cells exhibit substantially higher normalized natural frequencies, load-carrying capacities, and impact energy absorption compared to biomimetic honeycomb structures, underscoring the transformative potential of this innovative approach.

In addition to elucidating methodologies, the chapters provide practical insights into the implementation of machine learning techniques for structural optimization, equipping readers with a comprehensive understanding of the application landscape. While the bulk of the content delves into lightweight structural optimization, the techniques and methodologies expounded herein hold promise for extension into a myriad of other structural optimization domains, offering a roadmap for future exploration and innovation in the field.

Machine learning (ML) has rapidly transformed various industries by revolutionizing decision-making processes and unlocking insights from vast amounts of data. While traditionally associated with fields like engineering and computer science, ML has found extensive applications across diverse domains, including healthcare, finance, agriculture, transportation, and social sciences. To put machine learning in a larger and wider landscape, we would like to review and explore the multifaceted impact of ML in these other domains, highlighting some common key advancements, challenges, and opportunities.

8.2 MACHINE LEARNING APPLICATIONS IN OTHER AREAS OF STUDIES

8.2.1 HEALTHCARE APPLICATIONS

In the healthcare sector, machine learning (ML) has emerged as a powerful tool revolutionizing various aspects of patient care, disease diagnosis, treatment planning, and biomedical research. ML algorithms have shown remarkable capabilities in analyzing complex medical data, such as imaging scans, electronic health records (EHRs), and genomic sequences, to extract meaningful insights and improve clinical decision-making [1, 2].

One of the prominent applications of ML in healthcare is in medical imaging analysis, where deep learning algorithms have demonstrated exceptional performance in tasks such as lesion detection, tumor segmentation, and disease classification. For instance, convolutional neural networks (CNNs) have been successfully applied to mammography images for early breast cancer detection [3]. Similarly, deep learning models have shown promising results in detecting diabetic retinopathy from retinal images, aiding in timely intervention and prevention of vision loss [4].

ML-driven predictive analytics have also transformed patient care by enabling personalized medicine approaches. By integrating clinical data with genetic information and biomarkers, ML algorithms can predict patient outcomes, identify high-risk individuals, and recommend tailored treatment plans. For example, predictive models have been developed to forecast disease progression in patients with chronic conditions like diabetes and cardiovascular diseases [5, 6]. These models assist healthcare providers in stratifying patients based on their risk profiles and optimizing treatment strategies to improve outcomes and reduce healthcare costs [6].

Moreover, ML techniques play a crucial role in drug discovery and development, accelerating the identification of novel therapeutics and optimizing clinical trials. By analyzing large-scale biomedical datasets, including genomic, proteomic, and chemical data, ML algorithms can identify potential drug targets, predict drug efficacy, and optimize drug candidates' properties [7]. For instance, researchers have utilized deep learning models to predict the binding affinity between drug molecules and target proteins, facilitating the design of more effective drugs with reduced side effects [8].

Overall, ML applications in healthcare continue to advance rapidly, offering transformative opportunities to improve patient outcomes, enhance clinical decision-making, and drive biomedical innovation. However, challenges such as data privacy, regulatory compliance, and algorithmic bias must be addressed to ensure the ethical and responsible deployment of ML technologies in healthcare.

8.2.2 FINANCE AND BUSINESS APPLICATIONS

In the realm of finance and business, machine learning (ML) has become an indispensable tool for data-driven decision-making, risk management, and operational efficiency. ML algorithms are widely employed across various financial domains, including banking, insurance, investment management, and fintech, to analyze large volumes of data, identify patterns, and make predictions [9–11].

One of the primary applications of ML in finance is in fraud detection and risk assessment. ML algorithms, particularly anomaly detection techniques and predictive models, play a crucial role in detecting fraudulent activities, such as credit card fraud, identity theft, and money laundering [12]. By analyzing transactional data in real time, ML systems can identify suspicious patterns and flag potentially fraudulent transactions for further investigation [13].

Moreover, ML-driven predictive analytics are extensively utilized in investment management and algorithmic trading, where automated trading strategies rely on ML models to analyze market data, forecast price movements, and execute trades at optimal times [11]. Reinforcement learning algorithms, such as deep Q-networks (DQNs) and actor–critic methods, have been applied to develop trading agents capable of learning optimal trading policies from historical market data [12]. These ML-driven trading systems aim to generate alpha and outperform traditional investment strategies, leveraging the power of data-driven insights and computational techniques [13].

In addition to fraud detection and trading, ML techniques are also deployed in customer relationship management (CRM) and marketing analytics to improve customer segmentation, personalized recommendations, and campaign optimization [14]. By analyzing customer interactions, demographic data, and purchasing behavior, ML algorithms can identify valuable customer segments, predict individual preferences, and tailor marketing strategies to maximize engagement and conversion rates [15, 16].

Overall, ML applications in finance and business continue to evolve, driving innovation, efficiency, and competitiveness in the industry. However, challenges such as model interpretability, regulatory compliance, and ethical considerations must be addressed to ensure the responsible and sustainable deployment of ML technologies in financial services.

8.2.3 AGRICULTURAL AND ENVIRONMENTAL APPLICATIONS

The integration of machine learning (ML) technologies in agriculture and environmental science has revolutionized traditional farming practices, environmental monitoring, and conservation efforts. ML algorithms are increasingly applied across various agricultural domains, including precision agriculture, crop management, pest control, and soil health assessment, to enhance productivity, sustainability, and resilience [17–19].

Precision agriculture, in particular, has benefited significantly from ML-driven technologies, where sensor-based monitoring systems, satellite imagery, and IoT devices are leveraged to collect data on soil conditions, weather patterns, crop growth, and yield variability [20]. ML algorithms process and analyze these data streams to generate actionable insights, such as optimal planting schedules, irrigation strategies, and fertilizer applications, tailored to specific field conditions and crop requirements [21].

Furthermore, ML techniques are instrumental in crop disease detection and management, where early identification of plant diseases and pests is critical for mitigating yield losses and ensuring food security [22]. By analyzing multi-spectral images, hyperspectral data, and plant phenotypic traits, ML models can detect subtle signs of disease onset, classify disease types, and recommend targeted interventions, such as precision spraying or biological control methods [23].

Climate modeling and environmental forecasting are also areas where ML-driven approaches have made significant contributions, providing valuable insights into climate change impacts, extreme weather events, and ecosystem dynamics [24]. ML algorithms analyze historical climate data, satellite observations, and numerical simulations to develop predictive models for temperature trends, precipitation patterns, and natural disasters, aiding in risk assessment, adaptation planning, and policy formulation [25].

In summary, ML applications in agriculture and environmental science offer innovative solutions to address key challenges facing the agricultural sector, such as food security, resource sustainability, and climate resilience. However, efforts to democratize access to ML technologies, bridge the digital divide, and promote data sharing and collaboration are essential to ensure equitable and inclusive adoption of these technologies across diverse agricultural landscapes.

8.2.4 Transportation and Urban Planning

The transportation sector is undergoing a profound transformation with the integration of machine learning (ML) technologies, revolutionizing mobility services, infrastructure planning, and traffic management strategies. ML algorithms are deployed across various transportation domains, including autonomous vehicles, ride-sharing platforms, traffic prediction systems, and urban mobility solutions, to enhance safety, efficiency, and sustainability [26–28].

Autonomous vehicle technology represents one of the most prominent applications of ML in transportation, where ML algorithms enable vehicles to perceive their surroundings, navigate complex environments, and make real-time decisions autonomously [29]. Deep learning models, such as convolutional neural networks (CNNs) and recurrent neural networks (RNNs), process sensor data from cameras, LiDAR, and radar systems to detect objects, interpret traffic signs, and predict pedestrian behavior [30]. These ML-driven perception systems enable autonomous vehicles to operate safely and reliably in diverse driving conditions, accelerating the transition toward a future of autonomous mobility [31].

ML techniques are also instrumental in traffic prediction and congestion management, where predictive analytics models analyze historical traffic data, weather patterns, and event schedules to forecast traffic volumes, congestion hotspots, and travel times [28]. By leveraging machine learning algorithms, transportation agencies can optimize traffic signal timing, reroute traffic flow, and implement dynamic pricing strategies to alleviate congestion and improve overall network performance [29].

Furthermore, urban planning initiatives benefit from ML-driven simulations and optimization algorithms, enabling city planners to design sustainable transportation systems, prioritize infrastructure investments, and enhance accessibility for all residents [30]. ML-based tools, such as agent-based models and scenario planning frameworks, facilitate data-driven decision-making, stakeholder engagement, and policy evaluation in urban development projects [31].

In summary, ML applications in transportation and urban planning offer transformative opportunities to enhance mobility, reduce congestion, and improve the quality of life in cities and communities. However, challenges such as regulatory compliance, safety certification, and public acceptance must be addressed to realize the full potential of ML-driven transportation solutions in the real world.

8.2.5 SOCIAL SCIENCES AND HUMANITIES APPLICATIONS

Machine learning (ML) techniques are increasingly applied in social sciences and humanities research, enabling scholars to analyze vast amounts of textual and multimedia data and extract meaningful insights into human behavior, cultural trends, and historical phenomena [32, 33].

One of the primary applications of ML in the social sciences is in sentiment analysis and opinion mining, where ML algorithms analyze textual data from social media, news articles, and online forums to gauge public sentiment toward various topics, products, or events [34]. Natural language processing (NLP) techniques, including text classification, sentiment scoring, and topic modeling, enable researchers to identify trends, detect emerging issues, and understand public attitudes and perceptions [35].

Moreover, ML-driven recommender systems play a crucial role in content recommendation and personalized user experiences, particularly in the context of digital platforms and e-commerce websites [36]. Collaborative filtering algorithms, content-based filtering techniques, and hybrid recommendation approaches leverage ML models to analyze user preferences, behavior patterns, and item features to generate personalized recommendations tailored to individual user's interests and needs [37].

In addition to sentiment analysis and recommendation, ML techniques are also applied in historical research and cultural heritage preservation, where scholars leverage computational methods to analyze historical texts, manuscripts, and artifacts [38]. Optical character recognition (OCR) systems, named entity recognition (NER) algorithms, and topic modeling techniques enable researchers to digitize, transcribe, and analyze large corpora of historical documents, uncovering hidden patterns and insights into past events, social dynamics, and cultural practices [39].

Overall, ML applications in social sciences and humanities research offer innovative tools and methodologies to address complex research questions, analyze large-scale datasets, and gain deeper insights into human behavior, cultural evolution, and historical processes. However, interdisciplinary collaboration, ethical considerations, and methodological rigor are essential to ensure the responsible and meaningful use of ML technologies in social sciences and humanities research.

8.3   CHALLENGES, OPPORTUNITIES, AND PERSPECTIVES

While machine learning (ML) has demonstrated remarkable advancements across diverse domains, several challenges persist, including ethical considerations, data privacy concerns, and algorithmic biases.

Ethical issues related to data privacy, algorithmic fairness, and accountability require careful consideration to ensure responsible and ethical use of ML technologies [40–42]. The collection, storage, and use of personal data raise concerns about individual privacy rights, consent, and data security [43]. Moreover, ML algorithms may exhibit biases and discrimination, leading to unfair outcomes and reinforcing existing inequalities in society [44]. Addressing these ethical challenges requires interdisciplinary collaboration, regulatory oversight, and transparent governance mechanisms to promote ethical AI development and deployment [45].

Furthermore, challenges such as data quality, interpretability, and reproducibility pose significant obstacles to the adoption and deployment of ML technologies in real-world applications [46]. ML models rely on high-quality data to learn meaningful patterns and make accurate predictions, but data may be incomplete, biased, or noisy, leading to suboptimal performance and unreliable results [47]. Moreover, the black box nature of some ML algorithms hinders their interpretability and explainability, limiting their utility in critical decision-making contexts [48].

Despite these challenges, ML presents unprecedented opportunities for innovation, interdisciplinary collaboration, and societal impact across various domains [49]. By harnessing the power of data-driven insights, computational techniques, and predictive analytics, ML technologies have the potential to address complex societal challenges, drive economic growth, and improve the quality of life for individuals and communities [50]. However, realizing this potential requires collective efforts from researchers, policymakers, and industry stakeholders to promote responsible AI development, foster public trust, and ensure equitable access to AI technologies.

In summary, while machine learning offers transformative opportunities to advance knowledge, drive innovation, and address societal challenges, it also poses significant ethical, technical, and societal challenges that must be addressed to realize its full potential and ensure responsible and equitable deployment in diverse domains.

In the design and optimization of mechanical metamaterials such as lattice structures and thin-walled cellular structures, although machine learning is a new area of study, significant progress has been made in the past several years. We foresee that machine learning will become a powerful tool in discovering new mechanical metamaterials in the near future. Like other simulation tools such as finite element analysis, machine learning software packages will be in the design toolbox soon and will be available for engineers to design and optimize engineering structures and devices. Similar to the materials genome initiative (MGI) project, we recommend that some searchable databases consisting of structure and property information be established and be open to the public. We also suggest that an international organization be formed, and a conference be held regularly for scientists and engineers to share ideas. We are sure that when this book is available to readers, more progress should have been made in this area of study. Therefore, this book focuses more on fundamental knowledge and skills and serves as a stepstone for advanced studies. We hope that this book will attract new ideas and motivate more researchers to engage in this exciting area of study.

REFERENCES

1. Litjens, et al. A survey on deep learning in medical image analysis. Medical Image Analysis, 42:60–88, (2017).
2. Beam AL, Kohane IS. Big data and machine learning in health care. JAMA, 319(13):1317–1318, (2018).
3. Abdelhafiz D, Yang C, Ammar R, Nabavi S. Deep convolutional neural networks for mammography: Advances, challenges and applications. BMC Bioinformatics, 20:281, (2019).
4. Dai L, et al. A deep learning system for detecting diabetic retinopathy across the disease spectrum. Nature Communications, 12(1):3242, (2021).
5. Sheer R, et al. Predictive risk models to identify patients at high-risk for severe clinical outcomes with chronic kidney disease and type 2 diabetes. Journal of Primary Care Community Health, 13:215013192211063726, (2021).
6. Damen JA, et al. Prediction models for cardiovascular disease risk in the general population: Systematic review. British Publisher of Medical Journals, 353:i2416, (2016).
7. Vora LK, Gholap AD, Jetha K, Thakur RRS, Solanki HK, Chavda VP. Artificial intelligence in pharmaceutical technology and drug delivery design. Pharmaceutics, 15(7):1916, (2023).
8. Zhang H, Liu X, Cheng W, Wang T, Chen Y. Prediction of drug-target binding affinity based on deep learning models. Computers in Biology and Medicine, 174:108435, ISSN 0010-4825, (2024).
9. Pattnaik D, Ray S, Raman R. Applications of artificial intelligence and machine learning in the financial services industry: A bibliometric review. Heliyon, 10(1):e23492, ISSN 2405-8440, (2024).
10. Sargeant H. Algorithmic decision-making in financial services: Economic and normative outcomes in consumer credit. AI Ethics, 3:1295–1311, (2023).
11. Cao L, Yang Q, Yu PS. Data science and AI in FinTech: An overview. International Journal of Data Science, 12:81–99, (2021).
12. Waleed HS, Andrew G, John Y. Financial fraud: A review of anomaly detection techniques and recent advances. Expert Systems with Applications, 193:116429, (2022).
13. Galina B, Helmut K. Reducing false positives in fraud detection: Combining the red flag approach with process mining. International Journal of Accounting Information Systems, 31:1–16, (2018).
14. Rane N, Choudhary S, Rane J. Hyper-personalization for enhancing customer loyalty and satisfaction in customer relationship management (CRM) systems. Social Science Research Network, (2023).
15. Jisun A, Haewoon K, Soon, gyo J, Joni S, Bernard J. Customer segmentation using online platforms: Isolating behavioral and demographic segments for persona creation via aggregated user data. Social Networks, 8:54, (2018).

16. Andrés M, Claudia S, Sergiy P, Clemens P, Markus H. A machine learning framework for customer purchase prediction in the non-contractual setting. European Journal of Operational Research, 281(3):588–596, (2020).

17. Tawseef AS, Tabasum R, Faisal RL. Towards leveraging the role of machine learning and artificial intelligence in precision agriculture and smart farming. Computers and Electronics in Agriculture, 198:107119, (2022).

18. Benos L, Tagarakis AC, Dolias G, Berruto R, Kateris D, Bochtis D. Machine learning in agriculture: A comprehensive updated review. Sensors, 21(11):3758, (2021).

19. Mesías-Ruiz GA, Pérez-Ortiz M, Dorado J, De Castro AI, Peña JM. Boosting precision crop protection towards agriculture via machine learning and emerging technologies: A contextual review. Frontier of Plant Science, 14:1143326, (2023).

20. Popescu SM, Mansoor S, Wani OA, Kumar SS, Sharma V, Sharma A, Arya VM, Kirkham MB, Hou D, Bolan N, Chung YS. Artificial intelligence and IoT driven technologies for environmental pollution monitoring and management. Frontier of Environmental Science, 12:1336088, (2024).

21. Musanase C, Vodacek A, Hanyurwimfura D, Uwitonze A, Kabandana I. Data-driven analysis and machine learning-based crop and fertilizer recommendation system for revolutionizing farming practices. Agriculture, 13(11):2141, (2023).

22. Buja I, Sabella E, Monteduro AG, Chiriacò MS, De Bellis L, Luvisi A, Maruccio G. Advances in plant disease detection and monitoring: From traditional assays to in-field diagnostics. Sensors, 21(6):2129, (2021).

23. Anne-Katrin M. Plant disease detection by imaging sensors – Parallels and specific demands for precision agriculture and plant phenotyping. Plant Disease, 100(2):241–251, (2016).

24. Alexander YS, Bridget RS. How can big data and machine learning benefit environment and water management: A survey of methods, applications, and future directions. Environmental Research Letters, 14:073001, (2019).

25. Federica Z, Elisa F, Christian S, Silvia T, Sinem A, Andrea C, Antonio M. Exploring machine learning potential for climate change risk assessment. Earth-Science Reviews, 220:103752, (2021).

26. Nikitas A, Michalakopoulou K, Njoya ET, Karampatzakis D. Artificial intelligence, transport and the smart city: Definitions and dimensions of a new mobility era. Sustainability, 12(7):2789, (2020).

27. Haghighat AK, Ravichandra-Mouli V, Chakraborty P et al. Applications of deep learning in intelligent transportation systems. Journal of Big Data Analysis, 2:115–145 (2020).

28. Hasan U, Whyte A, Jassmi H. A review of the transformation of road transport systems: Are we ready for the next step in artificially intelligent sustainable transport? Applied System Innovation, 3(1):1, (2020).

29. Khayyam H, Javadi B, Jalili M, Jazar RN Artificial Intelligence and Internet of Things for Autonomous Vehicles. Nonlinear Approaches in Engineering Applications. Springer, (2020).

30. Fayyad J, Jaradat MA, Gruyer D, Najjaran H. Deep learning sensor fusion for autonomous vehicle perception and localization: A review. Sensors, 20(15):4220, (2020).

31. Arzoo M, Kumar N. Deep learning models for traffic flow prediction in autonomous vehicles: A review, solutions, and challenges. Vehicular Communications, 20:100184, ISSN 2214-2096, (2019).

32. Wagner-Pacifici R, Mohr JW, Breiger RL Ontologies, methodologies, and new uses of big data in the social and cultural sciences. Big Data & Society, 2(2), (2015).

33. Xiaoling S, Yiwan Y. Knowledge discovery: Methods from data mining and machine learning. Social Science Research, 110:102817, (2023).
34. Krishna PK. A literature review on application of sentiment analysis using machine learning techniques. International Journal of Applied Engineering and Management Letters, 4(2):41–77, (2020).
35. Gutierrez E, Karwowski W, Fiok K, Davahli MR, Liciaga T, Ahram T. Analysis of human behavior by mining textual data: Current research topics and analytical techniques. Symmetry, 13(7):1276, (2021).
36. Khrais L. Role of artificial intelligence in shaping consumer demand in e-commerce. Future Internet, 12(12):226, (2020).
37. Amin SA, Philips J, Tabrizi N Current Trends in Collaborative Filtering Recommendation Systems. Services, Springer, 11517, (2019).
38. Fontanella F, Colace M, Molinara A, Scotto DF, Stanco F. Pattern recognition and artificial intelligence techniques for cultural heritage. Pattern Recognition Letters, 138:23–29, (2020).
39. Memon J, Sami M, Khan RA, Uddin M. Handwritten optical character recognition (OCR). A Comprehensive Systematic Literature Review, 8:142642–142668, (2020).
40. Martin K. Ethical implications and accountability of algorithms. Journal of Bus Ethics, 160:835–850, (2019).
41. Tsamados A et al. The Ethics of Algorithms: Key Problems and Solutions. In: Floridi, L. (eds) Ethics, Governance, and Policies in Artificial Intelligence. Philosophical Studies Series, 144. Springer, (2021).
42. Mittelstadt BD, Allo P, Taddeo M, Wachter S, Floridi L The ethics of algorithms: Mapping the debate. Big Data & Society, 3(2), (2016).
43. Tony KT, Sudhir K. Privacy rights and data security: GDPR and personal data markets. Management Science, 69(8):4389–4412, (2022).
44. Pot M, Kieusseyan N, Prainsack B. Not all biases are bad: Equitable and inequitable biases in machine learning and radiology. Insights Imaging, 12:13, (2021).
45. Felzmann H, Fosch-Villaronga E, Lutz C et al. Towards transparency by design for artificial intelligence. Science & Engineering Ethics, 26:3333–3361, (2020).
46. Radanliev P, Santos O, Brandon-Jones A, Joinson A. Ethics and responsible AI deployment. Frontier of Artificial Intelligence, 7:1377011, (2024).
47. Studer S, Bui TB, Drescher C, Hanuschkin A, Winkler L, Peters S, Müller K-R. Towards CRISP-ML(q): A machine learning process model with quality assurance methodology. Machine Learning and Knowledge Extraction, 3(2):392–413, (2021).
48. Hassija V, Chamola V, Mahapatra A et al. Interpreting black-box models: A review on explainable artificial intelligence. Cognitive Computation, 16:45–74 (2024).
49. Yogesh KD, et.al, Artificial intelligence (AI): Multidisciplinary perspectives on emerging challenges, opportunities, and agenda for research, practice and policy. International Journal of Information Management, 57:101994, (2021).
50. Strielkowski W, Vlasov A, Selivanov K, Muraviev K, Shakhnov V. Prospects and challenges of the machine learning and data-driven methods for the predictive analysis of power systems: A review. Energies, 16:1–31, (2023).

Index